About Island Press

Island Press is the only nonprofit organization in the United States whose principal purpose is the publication of books on environmental issues and natural resource management. We provide solutions-oriented information to professionals, public officials, business and community leaders, and concerned citizens who are shaping responses to environmental problems.

In 2002, Island Press celebrates its eighteenth anniversary as the leading provider of timely and practical books that take a multidisciplinary approach to critical environmental concerns. Our growing list of titles reflects our commitment to bringing the best of an expanding body of literature to the environmental community throughout North America and the world.

Support for Island Press is provided by The Nathan Cummings Foundation, Geraldine R. Dodge Foundation, Doris Duke Charitable Foundation, Educational Foundation of America, The Charles Engelhard Foundation, The Ford Foundation, The George Gund Foundation, The Vira I. Heinz Endowment, The William and Flora Hewlett Foundation, Henry Luce Foundation, The John D. and Catherine T. MacArthur Foundation, The Andrew W. Mellon Foundation, The Moriah Fund, The Curtis and Edith Munson Foundation, National Fish and Wildlife Foundation, The New-Land Foundation, Oak Foundation, The Overbrook Foundation, The David and Lucile Packard Foundation, The Pew Charitable Trusts, The Rockefeller Foundation, The Winslow Foundation, and other generous donors.

The opinions expressed in this book are those of the author(s) and do not necessarily reflect the views of these foundations.

About the Society for Ecological Restoration

The Society for Ecological Restoration (SER) is an international, nonprofit organization. Its members are actively engaged in ecologically sensitive repair and management of damaged ecosystems through an unusually broad array of experience, knowledge sets, and cultural perspectives.

The mission of SER is to serve the growing field of ecological restoration by facilitating dialogue among restorationists, ecouraging research, promoting awareness of and public support for restoration and restorative management, contributing to the public policy discussions, and recognizing those who have made outstanding contributions to the field of restoration.

Society for Ecological Restoration, 1955 W. Grant Road, No. 150, Tucson, AZ 85745. Tel.: (520) 622-5485, Fax: (520) 622-5491, E-mail: info@ser.org.

Wildlife Restoration

SOCIETY FOR ECOLOGICAL RESTORATION

The Science and Practice of Ecological Restoration
A series edited by James Aronson and Donald A. Falk

Wildlife Restoration: Techniques for Habitat Analysis and Animal Monitoring, by Michael L. Morrison

Wildlife Restoration

TECHNIQUES FOR HABITAT ANALYSIS AND ANIMAL MONITORING

Michael L. Morrison

Foreword by
Paul R. Krausman

SOCIETY FOR ECOLOGICAL RESTORATION

ISLAND PRESS
Washington • Covelo • London

Library of Congress Cataloging-in-Publication Data
Morrison, Michael L.
  Wildlife restoration : techniques for habitat analysis and animal monitoring / Michael L. Morrison ; foreword by Paul R. Krausman.
     p. cm.—(The science and practice of ecological restoration)
Includes bibliographical references
  ISBN 1-55963-936-9 (cloth : alk. paper) — ISBN 1-55963-937-7 (pbk. : alk. paper)
1. Restoration ecology. 2. Habitat (Ecology) 3. Wildlife management. I. Title. II. Series.
  QH541.15.R45+   2002002903

British Cataloguing-in-Publication Data available.

Printed on recycled, acid-free paper ♲

Manufactured in the United States of America
09 08 07 06 05 04 03 02   8 7 6 5 4 3 2 1

CONTENTS

FOREWORD

It is appropriate for this series, "The Science and Practice of Ecological Restoration" initiated by the Society for Ecological Restoration (SER) and Island Press, to begin with a book on wildlife habitat restoration. SER's mission is to serve the growing field of ecological restoration by facilitating dialogue among restorationists, encouraging research, and promoting public support for restoration and restorative management. Members of this international group are actively engaged in ecologically sensitive repair and management of ecosystems in more than 35 nations. To be successful in this global endeavor, restorationists need to be well versed in the basics of ecology, the scientific study of the relationships between organisms and their environment.

Wildlife habitat around the world has been severely altered by the ax, plow, cow, fire, and gun. Ironically, as biologist Aldo Leopold observed nearly 70 years ago, these same tools can be used to restore altered habitats. For restoration to succeed in practice, however, our objectives must be clear and our methods must be appropriate, realistic, and lead to measurable results. Measurement is particularly important because it represents the link between restoration outcomes and management. This book offers an outline for the development of such measurement practice and provides a solid understanding of the conceptual and practical problems involved in maintaining, enhancing, and altering habitat for wild animals. This is important to restoration science and practice because, as Mike Morrison notes in the introduction, "much of restoration involves, directly or indirectly, improving conditions for native species of wildlife." Although wildlife populations are not always the focus of restoration efforts, most projects have at least some effect, intended or not, on animal populations.

Wildlife management in many parts of the world has evolved to encompass several main stages, including laws on hunting, predator control, reservation of land, artificial replenishment of animals, and environmental control. In North America, these stages have unfolded over the last two centuries and remain part of various management objectives. Today the emerging emphasis is on habitat restoration and management, which might be considered as the sixth stage in the evolution of the field. The Society for Ecological Restoration has developed in large part to support the efforts of those working to restore habitats for wild animals as well and plant and animal communities more generally.

As the science of wildlife management (including the management of related habitats) has developed, there has been a parallel emphasis on understanding the origins and conservation of global biodiversity. This trend is important for several reasons. First, the world's increasing human population continues to degrade the habitats that animals depend on, creating the imperative for their restoration. Second, ecological science is making rapid progress in understanding the distribution and origins of biological diversity—progress that can reveal new products and benefits for human use as well as alleviate environmental destruction. Finally, the destruction of natural animal habitats is causing an accelerating rate of extinctions, requiring a proactive stance toward restoring functioning habitats.

These three factors alone illustrate the importance of restoring wildlife habitats in order to enhance and maintain biodiversity as well as benefit human society. Without healthy ecosystems, the quality of all life, human and otherwise, suffers. Yet scientists have described fewer than 2 million of the estimated 5 to 30 million species on earth, and most higher levels of biological organization (communities and ecosystems) are still poorly characterized.

The primary causes of species endangerment in most parts of the world are anthropogenic: resource extraction, habitat destruction, spreading of nonnative species, urbanization, agriculture, outdoor recreation, unsustainable levels of livestock grazing, and others. To maintain the earth's biodiversity and minimize its decline, land managers and restorationists will have to shift the center of gravity in their philosophy from human uses to biological diversity management, conservation, and restoration. If biodiversity is to become an overriding management goal, forest land and rangeland cannot be managed as before. Instead, land (and water) management will be coordinated closely with habitat management strategies—primarily to enhance biodiversity and secondarily to sustain production of timber, livestock, and recreation opportunities.

To meet the challenges of this new era, in which society aims to restore habitats for wildlife and maintain biodiversity as a primary objective of management, new principles are needed. For example:

- *Cumulative effects.* Land managers and restorationists will need to assess the cumulative impact of individual projects on regional populations and resources beyond their control. This means emphasizing biotic integrity based on natural history and sound ecology, not politics. Multispecies management should replace single-species management, and management plans will have to promote diversity *among* habitats as well as within individual sites.
- *Ecological dynamics.* A broader view is needed even when working with individual species populations. Enough space with the appropriate habitat needs to be allocated for all aspects of an animal's life history, for example, including metapopulation dynamics, seasonal use and migration, sustainable habitat patch size, minimum viable populations, and all stages of succession required for survival.
- *Integration of ecology and planning.* Land managers and restorationists need to be informed and must learn to anticipate and correct problems. This requires not only a stream of accurate data but also a management context that addresses regional ecosystem dynamics as well as the status of individual sites and populations. Here we can see the importance of monitoring and measurement as Morrison stresses them, for they provide the critical feedback between actions and future decisions.

These principles are incorporated into the fabric of the new field of ecological restoration. Ironically, because of increasing human domination of the earth, many landscapes degraded by human actions are also dependent upon humans for their recovery. Despite our many social contradictions, we continue to find ways to maintain and restore habitats for all wildlife and to ensure the existence of healthy ecosystems. Restoration is a complex process, and the field of wildlife management has been a springboard for its development. Wildlife ecology is important to restoration and, as noted earlier, students of restoration ecology must be familiar with the basic principles that underlie the functioning of healthy populations, communities, and ecosystems. Mike Morrison's text is an excellent introduction to restoring wildlife and the habitats they depend on—one that builds a strong bridge between restoration ecology and the future of wildlife management.

PAUL R. KRAUSMAN
University of Arizona

SERIES INTRODUCTION

With this book we launch a new series: "The Science and Practice of Ecological Restoration," a joint effort of the Society for Ecological Restoration (SER) and Island Press. As the series title suggests, our aim is to create a new international forum devoted to advancing restoration science and practice, as well as their integration. Since this is the first book in the series, this is a good place to express our vision of the project and explain how it may relate to development of the restoration field.

Restoration is already emerging as a keystone discipline for ecosystem protection and management. Originally applied to just a limited domain of damaged sites, restoration is now a basic element of management in national parks and protected areas worldwide. Moreover, restoration is being applied at all levels—from local, community-based projects to strategies for global sustainability. And in coming decades, restoration as both a tool and a philosophy will continue to gain importance. We believe there are three fundamental reasons why restoration will become a dominant paradigm in the twenty-first century.

The first reason is the unfortunate reality that species, communities, and ecosystems worldwide continue to suffer accelerating decline at the hands of humanity. The major drivers of this emerging tragedy are by now familiar: a global human overpopulation barely under control; continual emergence of new and potentially destructive technologies that allow resource exploitation to reach into ever more remote regions; and a spectacular and unprecedented inequity in the distribution of resources among people—a world where 15 million starve to death each year and 50 percent of the world's people live at the edge of poverty and starvation while a fortunate minority, mostly living in affluent nations, enjoys wealth and luxury unimaginable to the other 5 billion inhabitants of the planet. The ecological consequences are equally familiar: massive topsoil loss; global climate change; rates of species extinction approaching the five great natural extinction events of the past 600 million years; and loss of indigenous cultures (hence loss of cultural diversity) approaching a terminal phase. All of these impacts appear to be converging in this new century. Under these circumstances, it is hardly surprising that people of conscience are trying to restore what they can, even as they try to stem the tidal wave of loss and reduce the cost of overdue restoration for future generations.

The second reason why people are turning to restoration is that it offers them a positive role in interactions with the world around them. In conventional approaches to conservation, people were often excluded from the frame entirely; there was no room for humans in "nature." (A later iteration of this philosophy was "Stay away but send money.") While attracting some people who could accept the idea of sharing the planet with other species, this philosophy also led critics to brand conservationists as elitists with no concern for the economic and recreational interests of the majority. In recent years, however, environmental protection has become a mainstream ethic (and socioeconomic policy) in its own right, taught from kindergarten up and supported—at least in word if not always in deed—by the majority of voters, taxpayers, and elected officials. Restoration takes this momentum a step further by inviting people *into* the frame with nature as healers: repairing trails, daylighting urban streams, burning prairies, turning old and unnecessary roads back into healthy habitat, rejuvenating battered wetlands, replanting damaged forests, even reintroducing lost native species. This phenomenon is not limited to citizens of the affluent nations. In his pioneering work in Guanacast Province, Costa Rica, biologist Dan Janzen has shown that local people may rally around the task of restoration when it becomes clear that their futures are at stake. Thus restoration gives us a chance to play the role of healer, not just destroyers or short-sighted exploiters. In the process, as restoration philosopher William Jordan has noted, it is we who are—or can be—equally restored.

The third reason why people are turning to restoration is simple: it works. No one can replace an extinct species, of course, and there will never be a substitute for original, large-scale ecosystems that have been left whole. Indeed you will find no more ardent supporters of conserving wild places than the community of restorationists. But restoration goes beyond conservation. If we take the notion of stewardship seriously, then restoration clearly has a huge role to play in coming decades.

Knowing how to heal is not a license for causing harm: this is the Hippocratic Oath that every doctor takes, and it has an exact analogue in restoration. But when serious ecological damage has been done, we will need a community of healers: people and institutions who have learned how to restore the health of populations and ecosystems. Restorationists know, perhaps better than anyone, how difficult the task of restoring communities and ecosystems really is. Rarely will you hear an expert in the field refer to the task of restoring as "easy." But restorationists are also amazingly creative. After all, they have learned their trade in the most trying circumstances imaginable: abandoned strip mines, eroded quarries and hillsides, polluted wetlands, channelized rivers, overgrazed and invaded grasslands, overgrown and clogged forests, battered desert canyons, dewatered streams and riparian areas, denuded tropical forests. Restorationists are learning, in short, to speak the language of nature in pain. Like any doctor, the restorationist does not "heal" the ecosystem: he or she creates conditions in which the patient can heal itself. In this sense, restoration offers ecological hope.

Along the way, restoration scientists and practitioners are gaining valuable knowledge about how populations, communities, and ecosystems work. And this brings us back to the motivation of this series: to compile a foundation of practical knowledge and scientific insight that will allow restoration to become the powerful healing tool that the world so clearly and desperately needs. Restoration science and practice are among the best tools we have for both local and global sustainability—a crucial consideration when we apply our work to areas of the world where restoring degraded ecosystems is not an amenity but literally a matter of life and death.

While Island Press has already published many outstanding books in environmental science and policy, this series should add some new dimensions. We intend this series to be truly international. SER and Island Press recognize that restoration is both a global need and a global endeavor, and we hope that this series will aid in this vital process at the widest possible scale. Faithful to the traditions of both partners, we intend to address terrestrial, insular, and aquatic ecosystems—at all levels of biology from population and organism biology to ecosystem studies, at all spatial scales from the glade to the landscape. We welcome social science as well as the natural and physical sciences, and we eagerly seek exceptional contributions from individual authors and collections of edited papers.

SER has a long tradition of bringing together land and resource managers, ecosystem designers and engineers, social scientists, and biologists.

Indeed, one of SER's fundamental objectives is to serve as an evolving forum for exchange of experimental results, observations, personal experiences, and hard-earned philosophical perspectives. For its part, Island Press continues to grow as the preeminent publisher of new ideas in environmental science, conservation, and policy, building along the way an equally interdisciplinary readership.

Our joint objective is to stimulate a series of foundation books worthy of the multidisciplinary, local-to-global nature of ecological restoration. We hope that restorationists, research ecologists, environmentalists, natural area managers, land and water resource policymakers, community leaders, environmental philosophers, and others will find value and use in this series. We believe that ecological restoration will become, as noted biologist and conservationist Edward O. Wilson has predicted, one of the keystones of ecology and environmental protection for the twenty-first century. We hope you agree.

JAMES ARONSON
Montpellier, France

DONALD A. FALK
Tucson, Arizona

Acknowledgments

I appreciate the assistance of many people in bringing this project to completion. I especially thank Don Falk for guidance throughout. John Rieger and Thomas Scott helped me get the project started by commenting on an early outline. Thomas Scott collaborated with me on Chapter 7. William Block and several anonymous referees provided valuable comments on an earlier draft of the text. Joyce VanDeWater did an excellent job of developing the graphics and artwork. The Zoological Society of San Diego, Bruce G. Marcot, Paul R. Krausman, Annalaura Averill-Murray, Suellen Lynn, and Thomas Scott donated photographs. I also thank Barbara Dean and the staff at Island Press for guiding me through the publication process.

MICHAEL L. MORRISON

INTRODUCTION

> The restorationist who sees himself . . . as the practical arm of
> some abstract, academic discipline is not likely to give great
> thought to the theoretical aspects of the restoration process. . . .
> He will probably view his failures as a result of personal igno-
> rance of the received wisdom. [Gilpin 1987:305]

Much of restoration involves improving the conditions for native species of wildlife. To be ultimately successful, our restoration plans must be guided by the needs of the wildlife in the project area. We need information on species abundances and distribution, both current and historic. We need details on habitat requirements, including proper plant species composition and structure. We need to understand niche relationships, especially constraints on resource acquisition. We need to know food requirements and breeding locations. We need to understand the role that succession will play in species turnovers. We need to know the problems associated with exotic species of plants and animals, the problems of restoring small, isolated areas, and more. In short: understanding wildlife is a complicated process that demands our careful consideration during all stages of restoration.

Moreover, the success of a restoration project should be judged by how wildlife species respond to it. Monitoring gives us feedback that allows us to modify the specific project and refine future projects. Restoration should include applications at all spatial scales—from broad-scale (landscape) projects, down to small, site-specific projects. Throughout, however, I emphasize an integrated ecosystem approach: a holistic approach.

This book provides ecologists, restorationists, administrators, and other professionals with a basic understanding of the fundamentals of wildlife populations and wildlife/habitat relationships. It covers the types of information you will need in planning as well as the basic tools you will need to

1

develop and implement a rigorous monitoring program. The primary monitoring themes covered here are experimental design and statistical analysis, including the sampling of rare species and populations. With this knowledge, restorationists will be equipped to discuss their needs with professional wildlife biologists. Although no special training or education is necessary, a knowledge of basic ecological concepts and basic statistics is helpful. The book addresses wildlife restoration by:

- Exploring the concepts of habitat and niche, their historic development, components, and spatial-temporal relationships, and their role in land management
- Explaining how wildlife populations are identified and counted
- Reviewing the practice of captive breeding, reintroduction, and translocation of animals
- Detailing techniques for measuring wildlife and wildlife habitat, including basic statistical techniques
- Discussing how wildlife and their habitat needs can be incorporated into restoration planning, especially concerning size of preserves, fragmentation, and corridors
- Outlining a holistic approach to the restoration of large landscapes (an integrated ecosystem approach)
- Examining how exotic species, competitors, predators, disease, and related factors influence restoration planning
- Developing a solid rationale for monitoring and good sampling design in restoration projects
- Analyzing the development and critique of individual monitoring projects
- Presenting case histories of restoration projects
- Pointing out further sources on wildlife/habitat relationships and monitoring

The book tackles the conceptual and practical problems involved in sampling wildlife populations and explains what wildlife biologists can, and cannot, achieve. I do not take a cookbook approach. Applying general prescriptions most often leads to unpredictable results, some of which may cause more harm than good (such as attracting unwanted exotic species). This book simply presents the basic tools you will need to *understand* ecological concepts that can be used to design restoration projects with specific goals for wildlife.

In this book you will learn the fundamental principles and practices for evaluating wildlife present in an area and determining their relationship with the habitat. Ecology is complicated. Thus there are many topics to be mastered: species lists, habitat use, ecological processes, monitoring, study design, statistical analyses, population processes, exotic species, disease, and parasites. Understanding these topics is essential if you wish to pursue restoration plans, endangered species recovery, population monitoring, impact assessment, reserve design, habitat conservation plans, and basic ecological relationships. Apart from covering these topics, this book indicates where you can find more advanced and detailed information.

It is not my purpose here to define the conditions—"historic conditions," for instance—that a restoration project should try to replicate. There is no end to debate over what the desired or "natural" time period should be—in effect, a debate over the *target* of restoration. My purpose is different: to help restorationists understand ecological processes as they relate to wildlife and their habitat. Nevertheless, there are basic ecological principles that suggest what is possible (and impossible) with regard to restoration. Just as there is no reason to undertake a research study that has no chance of success (because of time, funding, or logistical constraints), there is no reason to attempt to restore that which cannot be restored. Although I strongly support efforts to place restoration plans in the context of historic conditions (see Chapter 3), ecological reality must guide what we can and cannot achieve.

Generally I refrain from using the term "natural" or "natural community." Many restorationists think the baseline for natural conditions is a world without human impacts. Some take this notion further and include environmental impacts caused by the region's indigenous peoples under the rubric of "natural." Anderson (1996), for example, has depicted the ways in which an "ecosystem" could be viewed with and without various human impacts (Figure I.1). But those planning a restoration project must identify their own operating principles. In any event, achieving ecological conditions entirely free from human impacts is impossible due to local plant and animal extinctions, introduced species, migration and dispersal of plants and animals, and ecological processes. Developing a restoration plan requires a prioritization of goals based on knowledge of historic conditions, understanding of current regional conditions, knowledge of species-specific requirements, evaluation of legal requirements, and political reality. Regardless of its specific goals, every restoration project must consider the ecological processes operating in the area of concern.

The management of wildlife has recently been evolving along scientific, social, cultural, legal, ethical, and aesthetic dimensions. Traditionally wildlife

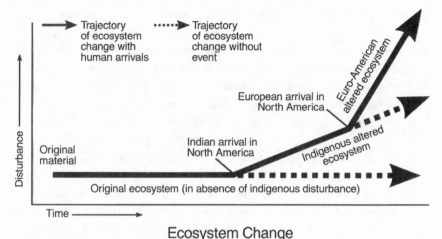

Ecosystem Change

FIGURE I.1.

Humans represent an ecological force that influences an ecosystem's trajectory. (From M. Kat Anderson, "Tending the Wilderness." *Ecological Restoration* 14[2]. Copyright 1996. Reprinted by permission of the University of Wisconsin Press.)

have been viewed solely as terrestrial vertebrates, especially game animals. Of late, however, the field of wildlife science has been expanding to include the conservation of all species of wildlife. (These issues are discussed in detail in Morrison et al. 1998:chap. 11.) In this book I adhere to the traditional concept of wildlife as terrestrial vertebrates. I do this not only because my own experiences are concentrated in this field of endeavor but also because I wish to keep this book focused and short.

For effective conservation and restoration, we must take an ecosystem approach. By this I mean consideration of a diversity of organisms along with their interactions with each other and with ecological conditions and processes. Understanding wildlife in an ecosystem context requires an understanding of:

• Population dynamics
• The evolutionary context of organisms, populations, and species
• Interactions between species that affect their persistence
• The influence of the environment on the vitality of organisms

An ecosystem context also demands an understanding of the role humans play in modifying environments, habitats, and wildlife populations (Grumbine 1994; Morrison et al. 1998:360).

Thus the restoration of wildlife and their habitats must be approached in a holistic way. In this manner we have a chance of understanding at least some of the factors that shape the distribution, abundance, and success of animals. Although we must certainly understand the broad landscape issues (such as population structure and dispersal corridors and mechanisms; see Bissonette 1997), I think the bottom-up approach to developing restoration plans will give us a more rigorous understanding of wildlife and result in the management practices needed to ensure success.

In summary, then, I hope this book fulfills at least two basic goals. First, I hope to give the restorationist a better understanding of the field of wildlife biology. Chapters 1 and 2 summarize the basic concepts. Second, I hope that wildlife biologists will learn a bit more about placing their studies in the context of restoration. Chapters 3, 7, and 8 should be especially useful in this regard.

References

Anderson, M. K. 1996. Tending the wilderness. *Restoration and Management Notes* 14:154–166.

Bissonette, J. A. (ed.). 1997. *Wildlife and Landscape Ecology: Effects of Pattern and Scale.* New York: Springer-Verlag.

Gilpin, M. E. 1987. Minimum viable populations: A restoration perspective. Pages 301–305 in W. R. Jordan III, M. E. Gilpin, and J. D. Aber (eds.), *Restoration Ecology: A Synthetic Approach to Ecological Research.* Cambridge: Cambridge University Press.

Grumbine, E. W. 1994. What is ecosystem management? *Conservation Biology* 8:27–38.

Morrison, M. L., B. G. Marcot, and R. W. Mannan. 1998. *Wildlife-Habitat Relationships: Concepts and Applications.* 2nd ed. Madison: University of Wisconsin Press.

CHAPTER 1

Populations

The ultimate goal of wildlife restoration is to ensure the survival and protection of individual animals. Management of habitat can create the conditions in which animals can maximize the number of viable offspring produced that in turn find mates and suitable environments and reproduce successfully. Fitness is influenced by the dynamics of interactions of individuals within a population, by interactions among populations and species, and by interactions between animals and their habitats and environments. To restore wildlife, therefore, requires knowledge of population dynamics and behavior. Successful restoration also requires that we understand the ecological processes that regulate population trends. Although habitat is essential to the survival of all species, by itself it does not guarantee long-term fitness and viability of population. In the Intermountain West of the United States, for example, macrohabitat conditions, measured as vegetation cover types and structural stages, for Townsend's big-eared bat (*Corynorhinus townsendii*) are estimated to have increased since the early 1800s by about 3 percent. Yet populations of this bat likely have *declined* over this period. Although the species uses a wide range of macrohabitats, substrates, and roosts, it is particularly vulnerable to human activity. Disturbing females with young adversely affects breeding success, and disturbing winter hibernacula may increase winter mortality (Nagorsen and Brigham 1993). In this case, therefore, the trend in macrohabitats belies the trend in populations—

even though providing such habitat is essential to species conservation and restoration (Morrison et al. 1998:49).

Let us begin by focusing on the spatial and geographic factors that influence habitats and environments, population structure, fitness of organisms, and, ultimately, the viability of populations. The restorationist must understand these parameters because they relate directly to the size and location of the area that needs to be restored to ensure survival of the species. It makes little sense, for example, to provide habitat for a species of interest if its long-term survival depends on immigration of new individuals and no allowance can be made for such immigration. Later in the chapter I review the topics of captive breeding, reintroduction, and translocation.

Population Concepts and Habitat Restoration

The traditional definition of a *population* is a collection of individuals of the same species that interbreed. Few wild creatures interbreed completely, however. Therefore, individuals of a species that have a high likelihood of interbreeding are called a *deme*. The term *subpopulation* is used to refer variously to a deme or to a portion of a population in a specific geographic location or as delineated by nonbiological criteria (such as administrative or political boundaries). Barriers to dispersal (water bodies, mountains, roads) and patchiness of resources prevent complete mixing of individuals and lead to heterogeneous distributions of individuals of a species.

Partial isolation of individuals and degrees of isolation among populations may result in metapopulation structures. A *metapopulation* occurs when "a species whose range is composed of more or less geographically isolated patches is interconnected through patterns of gene flow, extinction, and recolonization" and has been termed "a population of populations" (Lande and Barrowclough 1987:106). These component populations have been referred to as *subpopulations*. Metapopulations occur when environmental conditions and species characteristics provide for less than a complete interchange of reproductive individuals and there is greater demographic and reproductive interaction between individuals *within* than among subpopulations. Metapopulations occur frequently in wild animal populations. Metapopulations are held together by a multitude of factors, including dispersal and migration, habitat conditions, genetics, and behavior (Figure 1.1).

The structure of a population is a critical element of every restoration plan. If, for example, you are dealing with a metapopulation, your restoration plan needs to consider how far apart ample areas of habitat can be

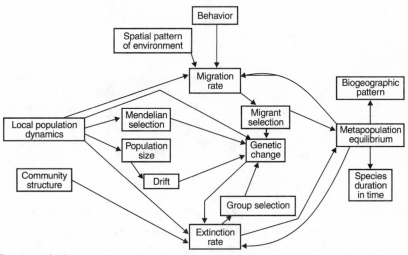

FIGURE 1.1.
Factors associated with metapopulation structure and dynamics. (From R. Levins, *Lecture on Mathematics in the Life Sciences,* Vol. 2, 75–107. Copyright 1970.)

located so that dispersal among the subpopulations may occur. Never forget that "habitat" is a species-specific concept (see Chapter 2). If the areas of habitat are too far apart, then extinction of the species within one location (subpopulation) might be permanent because there is no suitable area close enough to allow for dispersal and recolonization. Marsh and Trenham (2001) note, for example, that ponds should be considered as habitat patches ("ponds-as-patches") when you are planning for regional conservation of amphibians. This is not only a fundamental principle of wildlife ecology but also an essential component of a restoration plan.

Each subpopulation within a metapopulation may vary in abundance of animals. See, for example, the hypothetical but realistic example in Figure 1.2. Note the links between subpopulations: loss of a subpopulation that is located between other subpopulations could lead to further extirpations. In restoration planning one goal would be to identify the metapopulation structure and then locate restoration sites to enhance this structure—that is, to promote overall metapopulation persistence (the indicated subpopulation in the figure).

The rate at which animals disperse between subpopulation is related to the distance they must travel. The greater the distance between appropriate patches of habitat, the more difficulty an animal will have in locating the patch (and surviving long enough to find it). Information on dispersal abil-

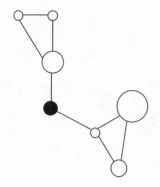

FIGURE 1.2.
Hypothetical arrangement and size of subpopulations (circles) in a metapopulation. Larger circle indicates a greater number of animals. Lines represent dispersal routes. The subpopulation indicated with a solid circle would be a key link between the other subpopulations—and a top-priority site for restoration.

ity is thus another critical component you will need in planning for animal restoration. A map of the distribution of bighorn sheep in southeastern California illustrates the complex metapopulation structure characteristics of wild animals (Figure 1.3). Note the loss of linkages between subpopulations caused by extirpation. Identifying formerly occupied locations is the first step in prioritizing your restoration efforts.

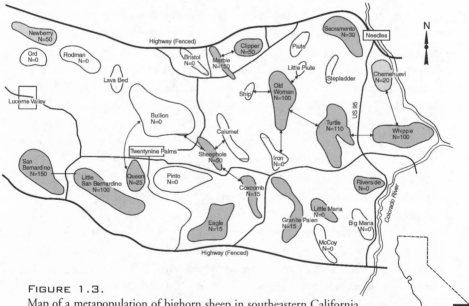

FIGURE 1.3.
Map of a metapopulation of bighorn sheep in southeastern California.
Stippled mountain ranges had resident populations of the size indicated. Mountain ranges with *N* = 0 are extirpated populations; ranges with no *N* value are not known to have had resident populations. Arrows indicate intermountain movements by sheep.
(From V. C. Bleich et al., *Conservation Biology* (4):383–390, Figure 1. Copyright 1990. Reprinted by permission of Blackwell Science, Inc.)

Population Dynamics and Viability

Population viability is the likelihood that a well-distributed population will persist to a specified future time period, typically a century or longer. The term *well-distributed population* refers to the need to ensure that individuals can freely interact where "natural" conditions once permitted. The time span for assessing viability should be scaled according to the species' life history, body size, longevity, and especially population generation time. Morrison et al. (1998:53) have suggested a rule of thumb: You should use at least 10 generations for gauging the lag effects of demographic dynamics and 50 generations for genetic dynamics. Longer time spans may be considered if environmental changes can be predicted beyond that time period. Thus for population of parrots with a generation time measured perhaps on the order of a decade, population viability should be projected over a century of demographic factors and five centuries for genetic factors. For a population of voles—with more rapid reproduction, shorter life spans, and far shorter generation times—viability may be projected only over a few years.

Population demography is the expression of a host of factors that influence individual fitness and population viability (Morrison et al. 1998:54). Vital rates may vary substantially in space and time as a function of food quality, weather, imbalance in sex ratios, and other factors. In some cases, populations respond to weather conditions, such as harsh winters, with lag effects measured in seasons or years. Morrison et al. (1998:59–62) review the influence of population genetics on conservation.

In population viability modeling, *thresholds* are conditions of the environment that, when changed slightly past certain values, cause populations to crash (Soulé 1980; Lande 1987). Such threshold conditions have given rise to the concept of *minimum viable populations* (MVPs) (Lacava and Hughes 1984; Gilpin and Soulé 1986). An MVP is the smallest population (typically measured in absolute number of individuals rather than their density or distribution) that can sustain itself over time and below which extinction is inevitable.

In the 1980s, researchers modeled MVPs by considering only genetic conditions of inbreeding depression and genetic drift. In theory, a minimum viable population is of a size (presumably the number of breeding individuals) below which the population is doomed to extinction and above which it is secure. One guideline proposed was the 50/500 Rule: populations of at least 50 breeding individuals should be maintained for ensuring short-term viability and 500 individuals for long-term viability (Gilpin and Soulé

1986). Such guidelines seldom pertain to real-world situations, however, and are based mostly on genetic considerations alone.

Viable population management goals often specify the need for large interconnected wildlife populations and diverse gene pools. Instead, your management goals should focus on understanding the natural conditions of a species in the wild. Sometimes populations are fully or partially isolated under natural conditions. In such cases, artificially inducing outbreeding among isolates—through captive breeding programs or manipulation of habitats, for example—may violate natural conditions.

The restorationist must carefully consider whether an area of given size can in fact support a viable population of a target species over the long term (however the time frame is defined). It makes little sense to provide habitat for a species if there is little probability of its long-term survival. It is, of course, very difficult to calculate the probability of a species' survival. At Naval Air Station, Lemoore (Kings County, California), for example, a population of the federally endangered Fresno kangaroo rat (*Dipodomys nitra-toides exilis*) has survived on a 40-ha area that has been completely isolated from any immigration for at least 20 years (Morrison et al. 1996). Nevertheless, your restoration plan should include at least a qualitative assessment of the likelihood that the species of interest will survive.

Metapopulations and Their Implications

The distribution, abundance, and dynamics of a population in a landscape are influenced by species attributes, habitat attributes, and other factors. *Species attributes* include movement and dispersal patterns, habitat specialization, demography (including density-dependence relations), and genetics of the populations. *Habitat attributes* include quality, size, spacing, connectivity, and fragmentation of habitat patches and the resulting availability and distribution of food, water, and cover. Other factors include a host of environmental conditions such as weather, hunting pressure, and influences from other species (Morrison et al. 1998:78).

Because of metapopulation structure, not all habitats are occupied simultaneously. This suggests that we should conserve habitats even if they seem unoccupied—and, in turn, that monitoring wildlife use of habitat should proceed for several years or season. Concluding that a species is absent where it is actually present is a Type II error (see Chapter 4) that can be corrected with adequate sample size and monitoring duration to increase the power of the statistical evaluation. The appropriate approach in designing a restoration plan depends on the size and fragility of the population and its habitats and on the project's objectives. Identifying the structure of the population of interest is critical if your restoration project is to establish or retain the population. The inherent metapopulation structure of many species emphasizes the need to examine the species' regional or landscape pattern of distribution.

Although it is generally not advisable to combine observations of habitat use patterns of individuals from different ecotypes, populations, geographic areas, or ecoregions (Ruggiero et al. 1988), synthesizing such information will at least give you a general understanding of a species' population structure. Such an analysis, based on literature review and expert opinion, will keep you from developing a restoration plan that is doomed to failure.

Distribution Patterns

The overall distribution and local abundance of many wildlife species are related in time and space. Many species have a "bull's-eye distribution" with their greatest area of abundance toward the middle of their overall range (and peripheral portions of the range in marginal conditions). Such distributions typically reflect several aspects of biophysical conditions and species ecology: the geographic range of suitable biophysical conditions; the species' range of tolerance of biophysical characteristics; and the occurrence of marginally suitable conditions at the periphery of the geographic range that may act as a sink habitat to hold nonreproductive individuals or individuals that have spread from higher-abundance areas during good reproductive years. (Sinks may also be habitat where mortality and emigration exceed natality and immigration.) Although many species have this bull's-eye pattern of distribution, the areas of highest density may not occur exactly in the center of the range. In studying censuses of birds from the North American Breeding Bird Survey (see Chapter 3), Brown et al. (1995) concluded that patterns of spatial and temporal variation in abundance should be considered when designing nature reserves and conserving biological diversity.

The bull's-eye distribution pattern is not always the rule, however. The abundance distribution of some species is truncated where biophysical conditions come to an abrupt halt—along mountain ranges, large rivers, or other major dispersal barriers, for instance, or at the edges of continents. An example is the wrentit (*Chamaea fasciata*), a species favoring chaparral, brush, and thickets, whose distribution truncates at its greatest density along the Pacific coast of the United States (Morrison et al. 1998:84). Thus, you cannot always assume that peripheral distribution of a species means marginal environmental conditions and lowest population densities.

Identifying areas of high animal density or areas of high environmental suitability—remembering that these are not always synonymous—can be important for management purposes. Wolf et al. (1996) found that releasing animals into the core of their historical range, and into habitat of high quality, were two factors contributing to success. Other factors included use of native game species, greater number of released animals, and an omnivorous diet. Many factors other than those listed here also contribute to population viability.

Here again the restorationist can improve the probability of success by surveying the literature and seeking expert opinion concerning the species' overall distribution relative to the restoration site. Is the site near the center of the range of the species? Or at the periphery? If it is near the center, colonization of the restoration site might be enhanced because immigration could occur from many directions and at relatively high frequency (depending on the species and the distance). But if the site is near the periphery of the range, colonization might be problematic. Again, the restorationist must consider these factors when designing the project.

Animal Movements

Movements of wildlife through their habitats impart particular dynamics to their populations. Movements especially important for habitat management include the following:

- *Dispersal:* one-way movement—typically of young away from natal areas
- *Migration:* a seasonal, cyclic movement—typically across latitudes or elevations to track resources or escape harsh seasonal conditions
- *Home range:* movement throughout a more or less definable and known space over the course of a day (or weeks or months) to locate resources
- *Eruption:* irregular movement into areas normally not occupied—a response to severe weather or sudden availability of high-quality resources

The use of such a movement classification system can aid habitat management by determining (1) which species are likely to occur in an area in a given season and thus the resources and habitats required during that season; (2) the number of species expected in an area over seasons and thus the collective resources and habitats required; and (3) the need to consider habitat conservation beyond the area of immediate interest. It may also aid in identifying habitat corridors used during movement—and thus habitats and geographic areas needing conservation focus (Figure 1.4). Morrison et al.

FIGURE 1.4.
The use of narrow strips of remnant vegetation as movement corridors by wildlife, as depicted here in British Columbia, needs much additional research. (Photo courtesy of Bruce G. Marcot.)

(1998:84–90) have reviewed the influence of animal movements on habitat management. Here I summarize these factors.

DISPERSAL. Dispersal—particularly of young away from natal areas—is an innate and genetically programmed adaptation to avoid inbreeding or resource competition and to locate mates. Often there is a wide variation in dispersal distances and patterns traveled among individuals even of the same cohort or family year; this may have adaptive significance in separating young and ensuring that at least some individuals reach suitable environments and locate mates. Dispersal patterns may vary by year, gender, time, and species interactions including competition.

One aspect of dispersal important for habitat management is recognition of dispersal barriers or strong filters. Few studies have empirically identified species-specific dispersal barriers or filters, although they could be critical in determining the occupancy of distributed habitat patches and persistence of low-density and small-size populations in the wild. Allen and Sargeant (1993), for example, found that a four-lane interstate highway altered dispersal directions of red foxes and apparently caused distant travel from natal areas.

MIGRATION. Migration is defined as the periodic (typically seasonal and annual) occurrence of animals in certain geographic locations and habitats. Migration of herding ungulates, large mammalian carnivores, some raptors, many Neotropical migratory songbirds, some amphibians, and other taxa can take place over short or long distances and latitudes. Long-distance migration may have arisen in long-term response to glacial fluxes and associated changes in climate and resource conditions (Chaney 1947), to avoid local competition for scarce resources, or owing to other factors, but responses are not well understood for most species. Most references concur that long-distance migration evolved in response to enormous seasonal differences in availability of food and other resources ("ultimate" factor); other factors such as competition, predator avoidance, and glaciation were associated but secondary ("proximate") factors.

HOME RANGE. How animals establish and use home ranges is important for managing habitat for populations. The restorationist should not just focus on home range sizes and amounts of habitat within the home range, however. There are other factors that could influence home range size and habitat selection, such as food supplies, density of conspecifics, effect of body size, competitors, predators, and landforms. In a sense, habitat selection is an optimization process that involves all of these factors and more—and home range is an expression of this process (Hall et al. 1997). This notion has important implications for management. Larger-than-average home ranges may or may not mean that insufficient habitat quality or amount has been provided (see Chapter 2).

ERUPTION. Some species undergo periodic eruptions in distribution and dispersal. Eruption refers to populations that suddenly exceed their normal boundaries or densities. (Irruption, by contrast, refers to invasion into a particular area.) In North America during periods of extreme northern winter conditions, some species typical of higher-latitude boreal forests, including the northern hawk owl (*Surnia ulula*), snowy owl (*Nyctea scandiaca*), gyrfalcon (*Falco rusticolus*), and white-winged crossbill (*Loxia leucoptera*), may occur intermittently further south in southern Canada and the northern United States. Such eruptive movements push the edge of population ranges into locations not normally occupied. More locally, seasonal eruptions or wandering movements occur with white-headed woodpeckers (*Picoides albolarvatus*), red crossbills (*Loxia currirostra*), redpolls (*Carduelis hornemanni* and *C. flammea*), evening grosbeaks (*Coccothraustes vespertinus*), and other species. Irregular conditions may help organisms colonizing new areas as founder populations. For instance, Scott (1994) reported that unusual prevailing winds aided an irruptive dispersal of black-shouldered kits (*Elanus caeruleus*) over 80 km of open ocean to San Clemente Island in southern California. Such movements may be important for determining potential colonization rates of peripheral habitats and the conservation value of habitat in the periphery of some species' range.

In restoration, the issue is to consider the population structure and movement patterns of each species of interest. Fortunately, natural history books provide at least general information on the patterns of dispersal and residency status of most species in a region.

Disturbed Habitats

Populations may respond to a perturbation either functionally or numerically. A functional response refers to changes in behavior—such as selecting different prey or using different substrates for resting or reproduction. Functional responses may also entail a temporary and localized increase in numbers resulting from immigration (or a decrease from emigration). A numerical response refers to absolute changes in abundance of individuals through changes in recruitment.

Disturbance of a habitat might elicit one or both responses. A crown fire in subalpine forests of the northern Rocky Mountains, for example, might increase suitability of habitat for black-backed woodpeckers (*Picoides arcticus*) in several ways: the fire may induce a temporary influx of foraging woodpeckers into the forest (a short-term functional response) as well as provide snag substrates for increased nesting density of woodpeckers in the area (a long-term numerical response). It is important to distinguish between such responses to understand whether your management activities—especially habitat restoration or enhancement—are truly increasing absolute population size (or simply redistributing animals). In some cases, simple redistribution may be the goal—such as warding off foraging waterfowl from grain fields and agricultural lands and into nearby wetland refuges. In other cases, redistribution and local increases—such as displacement from disturbed or fragmented habitat—may obscure an overall population decline.

Exotic Species

Intrusion by exotic species into natural environments has become a major challenge for habitat conservation and restoration (Coblentz 1990; Soulé 1990; OTA 1993). Exotic species come in all taxonomic groups and all environments. Exotic plants may disturb native ungulate use of rangelands (Trammell and Butler 1995) and handicap management of natural conditions in parks (Westman 1990; Tyser and Worley 1992). Exotic gamebird and big game introductions may affect distribution of native plants and animals (OTA 1993).

It is not always clear, however, which species are exotic (nonindigenous) and which are native. Some cases of range expansions may be natural; others may have been induced or enhanced by human alteration of environments; still others may have begun as a minor introduction by humans. An example of a "natural" invader of North America is the cattle egret (*Bubulcus ibis*), which spread from Africa to South American about 1880, reached Florida and Texas in the 1940s and 1950s, and rapidly expanded north and west in North American (Ehrlich et al. 1988). The brown-headed cowbird (*Molothrus ater*), which apparently evolved in the Great Plains, spread throughout most of North America during the 1900s—a movement likely resulting from clearing of forests and agricultural development (Morrison et al. 1999). Exotic escapee species that have spread throughout the continent include the European starling (*Sturnus vulgaris*): after two unsuccessful introductions, 60 birds were released into New York's Central Park in 1890—and within 60 years had spread to the Pacific and outcompeted and threatened many other bird species (Ehrlich et al. 1988).

Exotic species—or native species that have expanded into new regions—present special challenges for restorationists. The European starling, for example, will rapidly occupy natural cavities and artificial nest boxes that are meant for native species, thus dooming any attempt to restore native cavity-nesting birds. Although such eventualities are certainly frustrating, the restorationist should consider them when designing a wildlife restoration plan.

Three Roads to Recovery: Breeding, Reintroduction, and Translocation

Although habitat restoration is the primary focus of this book, in this section I review the use of captive breeding, reintroduction, and translocation in restoring animal populations. Throughout the world, these techniques have been used to assist with the restoration of numerous rare and endangered species. The goal of this section is to introduce readers to these three options for restoration programs.

The process of restoration often creates a system of subdivided populations of animals. Metapopulation biology, therefore, has direct application to the design and management of restoration projects. Metapopulation dynamics includes local population extinction, local population establishment or reestablishment, and movement or linkage among the various local populations. Metapopulation management, if properly applied, can reduce the probability of permanent extinction in local populations and help maintain genetic variability. Captive breeding facilities can maintain essentially fragmented populations. Thus captive propagation and restoration planning depend on a knowledge of metapopulation dynamics (Bowles and Whelan 1994).

Captive breeding and reintroduction are not the ideal means of achieving recovery of rare populations. Captive propagation is expensive, and reintroduction is problematic. Factors such as rates of gene flow among subpopulations, effective population size, mutation rates, and social structure must all be considered when planning restoration and reintroduction. Small populations might have been subject to past population bottlenecks, and genetic manipulation might be required to recover declining populations or to restore or maintain evolutionary potential. Maintenance of genetic variation and evolutionary potential are concerns for rare or isolated populations as well as captive populations. Captive breeding may lead to loss of genetic variation through random drift, to genetic adaptation through selection to the captive environment, and thus to inadequate adaptation for reintroduction to a restored location (Bowles and Whelan 1994).

Ramey et al. (2000) cite five key issues you must address when considering the augmentation of populations:

- Are there two lines of evidence (genetic, demographic, behavior) supporting the hypotheses that a severe population bottleneck has occurred?
- Would the introduction of additional animals degrade resource conditions, driving the wild animals to more rapid extinction?
- Was the population bottleneck due to a disease outbreak (or other specific occurrence) and can you eliminate the source of the problem?
- Are there habitat patches nearby to establish a population (or metapopulation) of larger size rather than a single, isolated population?

- How should the sex and age composition of an augmentation be structured?

These questions must be considered before you begin an animal restoration
project. Moreover, there is no reason to proceed with reintroduction if habitat and niche conditions are not appropriate.

Issues

Captive breeding and reintroduction have been successful and indeed may
be the only alternative to extinction in the wild. A restoration and conservation program that includes captive breeding and reintroduction involves
numerous overlapping steps (Figure 1.5). Here I review some of the issues
you should consider when planning for captive breeding, reintroduction, or

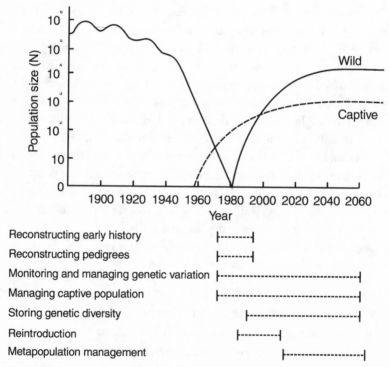

FIGURE 1.5.
Chronology of a captive breeding and restoration program. (From Mace et al. (eds.),
"Conserving Genetic Diversity with the Help of Biotechnology—Desert Antelopes as
an Example," Figure 1. Pages 123–134 in H. D. M. Moore et al. (eds.), *Biotechnology
and the Conservation of Genetic Diversity.* Copyright 1992, The Zoological Society of
London. (Reprinted by permission of Oxford University Press.)

translocation. I draw heavily from the summary of genetic considerations discussed by Lacy (1994).

GOALS. The goal of captive breeding programs is to support survival of the species, subspecies, or other defined unit in the wild. According to DeBoer (1992), meeting this goal requires:

- Propagating and managing highly endangered taxa, with prescribed levels of genetic diversity and demographic stability, for defined periods of time, to prevent extinction.
- Using captive programs as part of conservation strategies that manage captive and wild populations to ensure survival of these taxa in the wild—which means using captive populations to reestablish, reinforce, or re-create wild populations.

- Developing self-sustaining captive populations of rare or endangered taxa for education programs that benefit the survival of conspecifics in the wild (Figure 1.6).

FIGURE 1.6.
The Hawaiian goose, or nene, is being raised at the Zoological Society of San Diego's Keauhou Bird Center, Hawaii, for reintroduction into the wild. (Photo courtesy of Zoological Society of San Diego.)

Regardless of the species, there are certain prerequisites to meeting these goals:

• At the level of the individual, sufficient longevity and physical, physiological, and psychological well-being should be assured in the captive situation. This involves species-specific zootechnical, medical, and biological knowledge and research.
• At the level of breeding pairs and groups, sufficient reproduction should be assured to guarantee continuity over generations. This entails species-specific knowledge and research on reproductive biology, ethology, and related topics.
• At the level of population, the preservation of a genetic population structure should be assured that resembles the wild one as closely as possible.

GENETIC CONSIDERATIONS. Severe genetic problems (inbreeding, bottlenecks) may occur as a result of captive propagation, reintroduction, and translocation. These problems might not be readily apparent in the time scale of management programs, however, because few generations of vertebrates will have elapsed since the implementation of the program (Ramey et al. 2000).

There are two general types of genetic change in a captive population that have ramifications for restoration: *selection* may eliminate alleles that are maladaptive in the captive situation yet important for survival in the wild; *random genetic drift* may cause the cumulative loss of both adaptive and maladaptive alleles. The primary problem in captive propagation is that each successive generation is a sample of the previous generation. Thus the gene pool of the population that will eventually be reintroduced is invariably changed through captive generations. Rare alleles are especially susceptible to loss through genetic drift.

To preserve the genetics of animals that will be reintroduced, captive management must minimize adaptive and nonadaptive genetic changes. The effective population size necessary to minimize genetic changes in captive populations has been the subject of much debate. (See the reviews by Moore et al. 1992, Lacy 1994, and Gibbons et al. 1995.) The 50/500 rule calls for a minimum of 50 individuals for short-term breeding programs, but more than 500 individuals for long-term programs. The rationale for the long-term criterion is to allow new mutations to restore heterozygosity and additive genetic variance as rapidly as it is lost to random genetic drift. The concept of effective population size (Wright 1931) has several related meanings, including the number of individuals at which a genetically ideal population (one with random union of gametes) would drift at the rate of the observed population. The rate of genetic drift could be measured as the sampling variance of gene frequencies from parental to offspring generations. The primary goal of captive propagation, however, is to minimize all evolutionary (genetic) change, whether from random drift or selection to the captive environment. Another rule proposes that at least 90 percent of the genetic variation in the source (wild) population be maintained in the captive population (Lacy 1994). Ramey et al. (2000) suggest it is justifiable to intervene when there has been a severe genetic bottleneck—which they define as an effective population size of fewer than 10 individuals and a lack of gene flow with other outbred populations. Some reintroduction and translocation efforts, as we shall see, have been successful with as few as 10 individuals.

Genetic variation includes many related concepts—including genetically determined variation in morphology or behavior, variation in chromosomal structure, molecular variation in genes, allelic diversity, and heterozygosity of genes. The phenotype of an animal determines its physical properties and fitness. Thus quantitative genetic variation in phenotypes is of importance in designing genetic management for reintroduction. Much attention has focused, however, on theoretical models and prescriptions related to the management of underlying molecular genetic variation (that is, the presence of multiple genetic variation within a population; the mean or expected diversity per individual).

Heterozygosity can refer to diversity within individuals and diversity among individuals in a population. The proportion of loci for which the average individual is heterozygous is usually termed *observed heterozygosity.* The probability that two homologous genes randomly drawn from a population are distinct alleles is termed *expected heterozygosity*, or gene diversity. The 90 percent retention of genetic diversity guideline mentioned earlier refers to expected heterozygosity, and many management programs have used expected heterozygosity as an index of genetic variability. A population's capacity to adapt at all depends on the presence of sufficient variants, so allelic diversity might be critical to long-term persistence (Lacy 1994).

METAPOPULATION STRUCTURE. Most wild populations are divided into many subpopulations forming a metapopulation structure. Such a metapopulations structure can be used in captive breeding. Dispersing a captive population over a diverse environment may avoid directional selection that depletes genetic variation and promote selection that enhances genetic diversity. Dispersal also helps protect the population from epidemic disease and other catastrophes. Isolated or partly isolated populations tend to diverge genetically and thus lose different genetic variants. The metapopulation (as a whole) will therefore retain greater gene diversity (expected heterozygosity) and greater allelic diversity than would a single, large (panmictic) population (Lacy 1994). Management of metapopulation structure is, of course, difficult. Small subpopulations are subject to extinction and often require intensive management for extended periods of time (Mace et al. 1992).

Division of a population into several smaller populations comes with serious risks. Animals within each isolated population will become more inbred because of the limited mate choices and genetic drift. But if inbreeding depression was not severe, a large population reconstructed later by mixing animals from the isolated populations would be expected to have greater genetic variation and perhaps higher individual fitness than if the population had never been subdivided. Moving approximately one animal between populations per generation will usually prevent excessive inbreeding within populations—but at a cost of reduced effectiveness of the subdivided population structure in retaining variation overall. If five to ten animals are moved per generation, genetic divergence among subpopulations is largely prevented and the metapopulation is therefore equivalent to a panmictic population. Lacy (1994) concludes that perhaps the prudent approach to managing population structure is to mimic the amount of isolation typical of the wild population (or the best estimate of the characteristics of the wild population) before the onset of human-induced decimation and fragmentation.

Dividing or not dividing the population is not a mutually exclusive decision. In long-term breeding programs, occasional and temporary dividing is sometimes a tool for managing genetic diversity. And as noted earlier, there is seldom any justification for keeping all the captive animals in the same location.

Captive Breeding

Lacy (1994) has outlined three distinct phases of captive breeding. First, a captive program is established with wild-caught animals; this is the *founder phase*. Second, the captive population grows from the founders to the maximum size that can be supported by the program; this is the *growth phase*. Finally, the population is maintained at the captive equivalent of carrying capacity, and animals above carrying capacity are available for reintroduction; this is the *capacity phase*. The following paragraphs summarize the key points in each of these phases.

FOUNDER PHASE. The goal of a captive breeding program is to mimic the genetic composition of the wild population. Most captive programs, however, are begun as an effort of last resort after other plans (habitat restoration, for example, or removal of constraining factors such as predators) have failed. In such cases, much of the natural genetic variability has probably already been lost. Regardless of the size of the remaining wild population, attention should be given to taking a random genetic sample rather than focusing on capturing animals that are easy to obtain. But sampling based on genetics is not necessarily equivalent to randomly sampling individuals from across the remaining geographic range. To sample the genetic variability adequately, you need to know the structure of genetic variation in the wild. If such information is not available—which is often the case—then sampling can be based on observational data on population dynamics and social structure. If individuals are individually marked (color bands for birds, neck collars for large mammals), for example, then data should be available on lineages and efforts can be made to avoid oversampling from a specific bloodline. If animals are not individually marked or otherwise recognizable, then efforts should be made to avoid oversampling from individuals in isolated locations. (Unless dispersal of young is adequate, isolated groups of animals are likely to be closely related.)

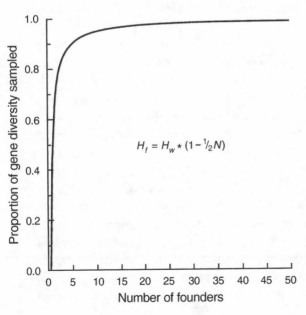

$$H_f = H_w * (1 - 1/2N)$$

FIGURE 1.7.
Proportion of the gene diversity (expected heterozygosity) present in a wild population (H_w) sampled in founders (H_f) drawn at random. (From R. C. Lacy, "Managing Genetic Diversity in Captive Populations of Animals," Figure 3.2. Pages 63–89 in M. L. Bowles and C. J. Whelan (eds.), *Restoration of Endangered Species: Conceptual Issues, Planning, and Implementation.* Copyright 1994, Cambridge University Press.)

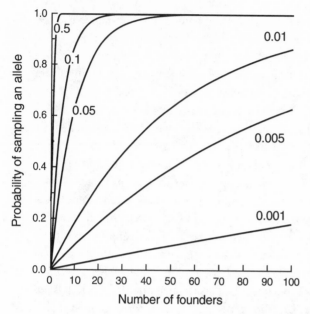

FIGURE 1.8.
Probability that an allele will be sampled at least once among founders drawn at random—for alleles of various frequencies (0.001, 0.005, 0.01, 0.05, 0.1, and 0.5) in the wild (source) population. (From R. C. Lacy, "Managing Genetic Diversity in Captive Populations of Animals," Figure 3.1. Pages 63–89 in M. L. Bowles and C. J. Whelan (eds.), *Restoration of Endangered Species: Conceptual Issues, Planning, and Implementation.* Copyright 1994, Cambridge University Press.)

To obtain most of the heterozygosity present in a wild population requires moderate numbers of founders. For example, obtaining at least 20 unrelated (randomly sampled) founders yields about 97 percent of the expected heterozygosity of the wild population because little additional heterozygosity is added with additional founders (Figure 1.7). A lack of genetic variation may prevent the establishment of a captive population capable of providing genetically diverse animals for reintroduction and thus impede recovery of the wild population. The probability that a particular allele will be included among the founder population, however, is determined by the frequency of the allele in the population. You might need a very large number of founders to get a high probability that an allele is included in the founder population. As depicted in Figure 1.8, it might take a large number of founders to achieve a high probability that rare alleles are included in a captive population (Lacy 1994).

Pedigree analysis is generally defined as the genetic study of a multigenerational population with ancestral links that are known or can be reasonably assumed or modeled. The emphasis is on examining the genetic structure imparted to the population by its pedigree relationships and evaluating the consequences of this structure on the long-term conservation of the population (Lacy et al. 1995). Current methods fall into three categories: the analytic calculation of genotype probabilities in completely known pedigrees;

the simulation of possible pedigrees from known aspects of a population's structure; and the determination of equations that broadly describe a population's genetic processes. A single gap in a pedigree may confound an analysis of genetically important individuals. Analyses may be restricted to those parts of the pedigrees that are completely known or to those populations with well-understood structure and dynamics. Moreover, assumptions can be made that allow analysis of uncertain data. Often several methods can be applied to increase the precision of results (Lacy et al. 1995).

GROWTH PHASE. Even if the initial founder population is a random sample from the wild population, it is likely that some founders will have only a few descendants and that specific alleles will be lost from the captive population. If founders produce disproportional numbers of offspring, then a conflict arises between the goal of balancing progeny and the goal of maximizing growth of the captive population. Ignoring a balance of contributors to the genetic diversity of the captive population will eventually lead to a decline in heterozygosity.

Because of inevitable changes in the genetic composition of the captive population, often there is pressure to begin releases into the wild before the initial goals of the captive program have been met. Releasing captive individuals while the population is in the growth phase is usually unwise. The argument that releasing the surplus progeny from prolific founders would not harm the captive population is weak in at least two ways. First, the ultimate goal is to restore the wild population, which should include both specific numerical and genetic objectives. Releasing surplus animals could further skew an already altered genetic composition in the wild animals. Second, a central conservation goal is to prevent extinction of the species of interest. Thus it would be wise to ensure the viability of the captive animals by using surplus individuals to establish another population whose survival is independent of the initial founder group (that is, establish another captive population). This second point makes sense even if many of the animals are never released. Indeed, they represent a fail-safe in case the central founder population sustains a catastrophic loss. Eliminating surplus individuals—or allowing them to be used in various behavioral or ecological test—is another option.

CAPACITY PHASE. The capacity phase is reached when the number of individuals with the ideal genetic composition is reached in captivity. The goal now is to maintain the health and genetic composition of the captive population while allowing for release of sufficient individuals (depending on the project) into the wild. Although alleles that are lost from the captive population cannot be replaced except by acquisition of more wild animals or by mutation, disparities in founder allele frequencies due to random drift or to selection (to the captive environment) can be partly reversed by giving priority breeding to the animals most likely to have unique or rare genes. That is: heterozygosity will be maximized if animals with the lowest overlap of genes are bred. Lacy (1994) has listed the many strategies for maximizing retention of genetic variation in captive breeding programs:

- Random mating
- Avoidance of close inbreeding (for example, half-sibling mating or closer)
- Circular mating schemes designed for maximum avoidance of inbreeding
- Equalization of family sizes
- Equalization of founder contributions
- Giving breeding priority to animals with the highest probability of carrying unique alleles
- Giving breeding priority to animals with the lowest mean kinship

Lacy concludes that no strategy outperforms a prioritization based on mean kinship, although comparable results could be obtained by selecting breeders with the highest probability of carrying unique alleles or by using a circular mating scheme designed for maximum avoidance of inbreeding. Regardless of the technique used, very close inbreeding should be avoided because of the high probability of inviable or infertile offspring (Moore et al. 1992; Kalinowski et al. 2000; Ramey et al. 2000).

Because the goal is to minimize losses of the founder population's genetic variation, breeding priority is given to those animals that descend from founders with the lowest representation in the descendant population. No completely satisfactory method exists for summarizing multiple contributions to an individual's genome into a single measure of founder value. Kinship, however, provides a genetic ranking that allows for optimal retention of heterozygosity (Lacy 1994).

Maximizing retention of heterozygosity usually, but not always, optimizes allelic diversity. Animals that have unique or rare alleles may also have very common alleles at other loci or at the homologous genes. Mating such animals could reduce heterozygosity because they share genes with much of the population; yet not breeding them could result in the loss of rare alleles. It is possible to manage for retention of rare alleles in two ways: by using simulation methods to determine the proportion of genes in each individual that are unique or rare in a population or by using exact methods of analysis to determine the probability of unique genes. You can avoid this conflict between preserving allelic diversity and preserving heterozygosity by pairing genetically valuable males with genetically valuable females so that descendants will not contain combinations of rare and common founder alleles. Perhaps the best overall strategy is to mate animals with the lowest mean kinship, to give somewhat higher priority to animals with an unusually high probability of having rare or unique alleles, and to avoid pairings among very close relatives (Lacy 1994). Of course there are numerous problems—such as social incompatibility among pairs and accidental deaths—that may prevent a strategy from being fully realized.

Mating animals with the lowest mean kinship will favor continued breeding by a parental generation over replacement of aging breeders by the offspring generation. Because offspring are genetic subsets of their parents, they can never have lower mean kinship to the population than do both parents. (That is: an offspring's kinship to each animal other than its parents is equal to the mean kinship of the parents to that other animal, and an offspring is related by at least one-half to each parent, while the parents might not be related to each other.) Replacing animals as breeders only when they die or cannot reproduce has several genetic advantages. Genetic drift occurs at the transition of generations, and alleles may be lost from a breeding population only when the breeders are retired or die. The ideal genetic management would be to immortalize the original wild-caught founders (Lacy 1994).

Reintroduction

Returning animals to locations that were formerly inhabited is one means of speeding restoration of an area. There are numerous components involved with successful reintroductions, however, including genetics of the animals and the suitability of the site (habitat conditions, presence of predators or competitors). In addition to using animals bred in captive situations, moving wild animals from one location to another (translocation) is a common technique. In this section I explore some of the techniques that have been used to reintroduce animals as part of a restoration effort.

CHARACTERISTICS OF THE CAPTIVE POPULATION. The choice of specific individuals for reintroduction is a key component of a successful animal restoration program. Given that mortality is high among reintroduced animals,

TABLE 1.1. Factors Considered Critical to Success of Woodland Caribou Restoration at Three Sites in the Western Lake Superior Region

	PREDATORS		WHITE-TAILED DEER	
Site	Wolves per 1000 km²	Black bears	Density	Incidence of brainworm
A	15	Pending	Low	44–60%
B	30	Common	High	> 90%
C	20	Absent	Absent	—

Source: From P. J. P. Gogan and J. F. Cochrane, "Restoration of Wildlife Caribou to the Lake Superior Region," Table 9.3. Page 235 in M. L. Bowles and C. J. Whelan (eds.), *Restoration of Endangered Species: Conceptual Issues, Planning, and Implementation*. Copyright 1994. Reprinted with permission of Cambridge University Press.

only animals that are surplus to the genetic needs of the captive population should be released. This is why, as noted earlier, the genetic composition of the captive population deserves so much attention. In practical terms, this means that animals descending from the most prolific lineages in captivity are the first to be released. Once these reintroduced individuals survive at an acceptable level, the genetic composition of the restored (or augmented) population can be diversified by the descendants of other lineages (Lacy 1994).

EVALUATION OF REINTRODUCTION SITES. The initial success of the reintroduction program depends, of course, on the survival of the released animals. In the long term, however, the released animals must not only survive but produce viable offspring. Thus the success of the program is enhanced when animals are released into high-quality habitat. The chapters that follow, especially Chapter 2, discuss the factors leading to successful survival and reproduction.

Resources central to survival and reproduction must be available at the release site. Thus a key step in the reintroduction program is to identify the critical factors and summarize their status at the release location. Table 1.1 shows a simple site evaluation guide for woodland caribou (*Rangifer tarandus caribou*). Note that the guide will include not only key habitat components (habitat quality) but also constraints on the use of these components, such as predators (wolf density) and competitors (deer density).

Translocation

Translocation is the relocation by humans of wild animals from one geographic site to another. Primarily it is used to establish a new subpopulation or enhance an existing subpopulation (by increasing numbers or manipulating genetic composition). Translocation is often used, with varying success, to conserve ungulate populations. For native game species, 20 to 40 founding animals have been found to be a sufficient number for translocation success (Gogan and Cochrane 1994). The probability of success is based, of course, on the condition of the habitat at the new site. (See Chapter 2 for details.) Not only must the physical environment and general habitat conditions be appropriate, but resources must be of adequate quantity and quality—and be available to the translocated animals. Thus careful analysis of the constraints, both real and potential (predators, competitors, human disturbance), on access to resources must precede a translocation program. Most of this discussion applies equally to the release of animals from a captive breeding program.

Because of the multitude of methods available (traps, snares, nets, immobilization drugs), I will not discuss capture techniques in this section. Many fine discussions of techniques for specific species are available (Bookhout 1994, Heyer et al. 1994; Wilson et al. 1996; see also Chapter 7). Once the animal is captured, it is essential that its time in captivity be minimized. Different species—and different individuals within a species—react differently to capture, transport, and handling. We can be certain, however, that captivity results in both behavioral and physiological stress on the individual. Here again, techniques for minimizing stress and maximizing survival are species-specific. Most translocation programs use veterinarians who are trained in the management of stress for the species in question.

There are two basic techniques for releasing animals: *soft release* and *hard release*. In soft releases, captured animals are held in captivity for an extended period of time (days to months) for a variety of behavioral and physiological reasons. This captivity may be in a laboratory or another holding location; or, more commonly, it is in a confined (caged, fenced) location at the eventual release site or nearby. The rationale for a soft release is that animals become accustomed to the release-site environment and observers can monitor animal condition prior to release. Naturally, food and other requirements must be provided, which increases animal contact with humans. Thus care must be taken to avoid habituating the animals to humans during extended soft releases. Soft releases are frequently used in captive breeding programs. For example, swift foxes (*Vulpes velox*) in Canada were paired and

TABLE 1.2. Survival of Swift Foxes Using Soft and Hard Releases

	Number	SURVIVAL		
Release method	radio-collared	to 6 mo.	to 12 mo.	to 24 mo.
Soft	45	55%	31%	13%
Hard	155	34%	17%	12%

Source: From L. N. Carbyn et al., "The Swift Fox Reintroduction Program in Canada from 1983 to 1992," Table 10.2. In M. L. Bowles and C. J. Whelan (eds.), *Restoration of Endangered Species: Conceptual Issues, Planning, and Implementation.* Copyright 1994. (Reprinted with permission of Cambridge University Press.)

held in field release pens (3.7 m × 7.3 m) in prairie environments for 1 to 8 months. They were placed in the pens in October or November and held during the mating season (January or February). If they did not produce off-spring, they were released the following spring. If they produced young, they and their young were released in summer or early fall (Carbyn et al. 1994).

In hard releases, animals are transported from the capture site (or captive rearing site) and released into the wild without any prior conditioning. The rationale for a hard release is to eliminate the stress that might accompany captivity at the new site. Returning to the swift fox example, animals were also released directly into the field without being placed in the release pens. Although soft-released animals initially had higher survival, no difference in survival was noted after 24 months (Table 1.2). Because hard releases were successful and cost-effective (no expense was incurred for condition in the field), the Canadian swift fox program continued using only hard releases. Carbyn et al. (1994) have emphasized that to reintroduce the swift fox it was necessary to determine whether its ecological niche was still present and if a minimum viable population could be established. The success of a restoration program—whether or not reintroductions or translocations are involved—rests fundamentally on the condition of the species' habitat and niche.

Another example of using translocation to enhance the status of a declining species is that of the mountain sheep (*Ovis canadensis*) in California. Mountain sheep have a naturally fragmented distribution. Many mountain sheep populations have been extirpated from historical ranges as a result of activities associated with human presence, including disease contracted from livestock, habitat destruction, and possible overharvest (Thompson et al.

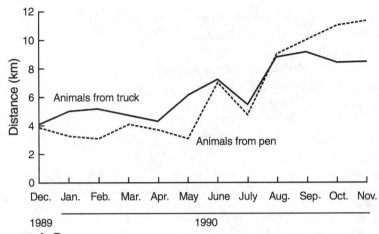

FIGURE 1.9.
Average distance from release site (an index of dispersal), by month, for animals transported by helicopter and held in a pen prior to release (dashed line) and for animals
released directly from a vehicle (solid line) in the Chuckwalla Mountains, Riverside
County, California, 1989 to 1990. (From Thompson et al., "Translocation Techniques
for Mountain Sheep: Does the Method Matter?" Figure 1. *Southwestern Naturalist*
46:87–93. Copyright 2001.)

2001). Mainly two techniques have been used to release mountain sheep: *direct (hard) release* on the periphery of a mountain range following vehicular transport from the capture site and *vehicular transport* followed by helicopter transport and holding animals in a temporary enclosure in the interior of a mountain range for a short time (6 to 8 hours) before release. Using an index of dispersal of animals from the release site, Thompson et al. (2001) found no statistically significant differences between the two methods (Figure 1.9). But 70 percent of the direct-release animals and only 30 percent of the penned animals had survived approximately one year following release. Thompson and colleagues concluded that the increased handling time required for penned animals (plus exposure to helicopter noise) jeopardized their survival.

Methods for releasing animals—regardless of the amount of time in captivity—are still in the development stage. No method is guaranteed to perform better than another, even within the same or closely related species. This level of uncertainty is due largely to the unique environmental situations confronting each program. Research into release techniques for passerine birds has received little attention with respect to large mammals. The

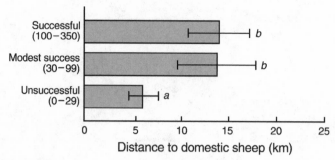

FIGURE 1.10.
Success of bighorn sheep translocations in relation to distance to the nearest domestic sheep, western United States, 1923–1997. Different letters denote statistical significance ($P < 0.05$). (From Singer et al., "Translocations as a Tool for Restoring Populations of Big Horn Sheep," Figure 1. *Restoration Ecology* 8:6–13. Copyright 2000. Reprinted by permission of Blackwell Science, Inc.)

captive breeding and release program for the endangered San Clemente loggerhead shrike (*Lanius ludovicianus mearnsi*), for example, is pioneering methods for passerines (Morrison et al. 1995). Numerous techniques are being tested: releasing individual adults, pairing of adults prior to release, releasing a group of captive-reared siblings simultaneously, and more. The survival of released birds, however, appears to be largely dependent on control of an exotic predator (feral cats)—indicating the critical need to evaluate the niche available to released birds. Likewise, releases of the endangered Hawaiian crow (*Corvus hawaiiensis*) are being inhibited by attacks on the crow by the Hawaiian hawk (*Buteo solitarius*)—itself a federally listed endangered species. When Singer et al. (2000) evaluated the success of 100 translocation attempts of bighorn sheep in six western states between 1923 and 1997, they classified 30 attempts as unsuccessful, 29 as moderately successful, and 41 as successful. Translocations were less successful when domestic sheep were located within 6 km of the bighorn sheep's use area (Figure 1.10).

Mortality due to predation is a primary cause of failure in reintroduction and translocation. Animals that have been isolated from predators, either throughout their lifetime in captivity or over evolutionary time (via translocations), might no longer express appropriate antipredator behavior. When Griffin et al. (2000) reviewed efforts to train captive-bred or translocated animals to avoid predators, they concluded that prerelease training may enhance the expression of antipredator behavior. Training techniques involve classical conditioning procedures whereby animals learn that models of predators are predictors of aversive events. Each technique is accompanied by potentially serious problems, however, that must be avoided through study and planning (Griffin et al. 2000:table 1).

Status and Future Prospects

Reintroduction and translocation are gaining increasing attention—and increasing use—throughout the world as a means of restoring and managing animals. Reintroductions involve all vertebrate groups. For example, rock iguanas (*Iguana pinguis*) were relocated between islands in the West Indies to establish a second population to serve as a reservoir for this endangered species. Although the translocation involved only eight individuals, it resulted in a successful new population (Goodyear and Lazell 1994). Helmeted honeyeaters (*Lichenostomus melanops cassidix*), a highly endangered species, are being reintroduced in southeastern Australia (Pearce and Lindenmayer 1998). In North America, translocation of the gray wolf (*Canis lupus*) has met with opposition in many places, although the translocations have generally been successful (Fritts et al. 1997). As the literature on evaluations and improvements in reintroduction and translocation grows (Griffith et al. 1989; Trulio 1995; Hein 1997), use of these techniques is likely to increase as a valuable restoration tool. Concurrently we must continue to develop and evaluate techniques for genetic management (Lacy et al. 1995), especially those applied to metapopulations composed of small subpopulations.

Lessons

This chapter has highlighted the critical role that population structure plays in the success of a restoration project for wildlife. No project should proceed without an assessment of the local and regional distribution of the animal species of interest. If time or funding constraints prevent a field assessment of the populations, then a thorough literature review, combined with the opinions of experts on the species and population ecology in general, should be undertaken to prevent the pursuit of a project whose goals are unattainable. At the very least this information will alert you to the likelihood that the target animals will colonize the site. Moreover, it should lead to a clear understanding of the difference between simple occupancy and successful survival and reproduction within the project area.

Most wild populations are distributed in a metapopulation structure. This is understandable given that, on a broad landscape scale, the habitat required by a species is distributed in patches. Understanding how a specific restoration site or patch relates to other patches is a critical step in restoration planning. Ultimately there is no reason to try to restore a site for a species if there is little likelihood that a new subpopulation could colonize the location. Likewise, there is no reason to attempt a translocation or reintroduction if the restoration site is too distant from another subpopulation. Again, a landscape perspective is necessary.

Captive breeding, reintroduction, and translocation are all viable methods of establishing (or reestablishing) an animal population at a restoration site. But first you must evaluate the genetic composition of wild and captive populations to ensure that a new population is suitable for a location. Although there are many success stories, there are even more failures. The field of captive breeding and reintroduction is growing rapidly, but few species have been studied at this time. As we shall see in the next chapter, detailed study is necessary to determine whether the species' habitat and niche requirements have been met before animals are brought to a site.

References

Allen, S. H., and A. B. Sargeant. 1993. Dispersal patterns of red foxes relative to population density. *Journal of Wildlife Management* 57(3):526–533.

Ballou, J. D., M. E. Gilpin, and T. J. Foose (eds.). 1995. *Population Management for Survival and Recovery: Analytical Methods and Strategies in Small Population Conservation.* New York: Columbia University Press.

Bleich, V. C., J. D. Wehausen, and S. A. Holl. 1990. Desert-dwelling mountain sheep: Conservation implications of a naturally fragmented distribution. *Conservation Biology* 4:383–390.

Bookhout, T. A. (ed.). 1994. *Research and Management Techniques for Wildlife and Habitats.* 5th ed. Bethesda, Md.: Wildlife Society.

Bowles, M. L., and C. J. Whelan. 1994. Conceptual issues in restoration ecology. Pages 1–7 in M. L. Bowles and C. J. Whelan (eds.), *Restoration of Endangered Species: Conceptual Issues, Planning, and Implementation.* Cambridge: Cambridge University Press.

Brown, J. H., D. W. Mehlman, and G. C. Stevens. 1995. Spatial variation in abundance. *Ecology* 76:2028–2043.

Carbyn, L. N., H. J. Armbruster, and C. Mamo. 1994. The swift fox reintroduction program in Canada from 1983 to 1992. Pages 247–271 in M. L. Bowles and C. J. Whelan (eds.), *Restoration of Endangered Species: Conceptual*

Issues, Planning, and Implementation. Cambridge: Cambridge University Press.

Chaney, R. W. 1947. Tertiary centers and migratory routes. *Ecological Monographs* 17:139–148.

Coblentz, B. E. 1990. Exotic organisms: A dilemma for conservation biology. *Conservation Biology* 4:261–265.

DeBoer, L.E.M. 1992. Current status of captive breeding programmes. Pages 5–16 in H.D.M. Moore, W. V. Holt, and G. M. Mace (eds.), *Biotechnology and the Conservation of Genetic Diversity.* Symposia of the Zoological Society of London, No. 64. Oxford: Oxford University Press.

Ehrlich, P. R., D. S. Dobkin, and D. Wheye. 1988. *The Birder's Handbook.* New York: Simon & Schuster.

Fritts, S. H., E. E. Bangs, J. A. Fontaine, M. R. Johnson, M. K. Phillips, E. D. Koch, and J. R. Gunson. 1997. Planning and implementing a reintroduction of wolves to Yellowstone National Park and central Idaho. *Restoration Ecology* 5:7–27.

Gibbons, E. F., Jr., B. S. Durrant, and J. Demarest. 1995. *Conservation of Endangered Species in Captivity: An Interdisciplinary Approach.* Albany: State University of New York Press.

Gilpin, M. E., and M. E. Soulé. 1986. Minimum viable populations: Processes of species extinction. Pages 19–34 in M. E. Soulé (ed.), *Conservation Biology: The Science of Scarcity and Diversity.* Sunderland, Mass.: Sinauer Associates.

Gogan, P.J.P., and J. F. Cochrane. 1994. Restoration of woodland caribou to the Lake Superior region. Pages 219–242 in M. L. Bowles and C. J. Whelan (eds.), *Restoration of Endangered Species: Conceptual Issues, Planning, and Implementation.* Cambridge: Cambridge University Press.

Goodyear, N. C., and J. Lazell. 1994. Status of a relocated population of endangered *Iguana pinguis* on Guana Island, British Virgin Islands. *Restoration Ecology* 2:43–50.

Griffin, A. S., D. T. Blumstein, and C. S. Evans. 2000. Training captive-bred or translocated animals to avoid predators. *Conservation Biology* 14:1317–1326.

Griffith, B. J., M. Scott, J. W. Carpenter, and C. Reed. 1989. Translocation as a species conservation tool: Status and strategy. *Science* 245:477–480.

Hall, L. S., P. R., Krausman, and M. L. Morrison. 1997. The habitat concept and a plea for standard terminology. *Wildlife Society Bulletin* 25:173–182.

Hein, E. W. 1997. Improving translocation programs. *Conservation Biology* 11:1270–1271.

Heyer, W. R., M. A. Donnelly, R. W. McDiarmid, L. C. Hayek, and M. S. Foster (eds.). 1996. *Measuring and Monitoring Biological Diversity: Standard Methods for Amphibians.* Washington, D.C.: Smithsonian.

Kalinowski, S. T., P. W. Hedrick, and P. S. Miller. 2000. Inbreeding depression in the Speke's gazelle captive breeding program. *Conservation Biology* 14:1375–1384.

Lacava, J., and J. Hughes. 1984. Determining minimum viable population levels. *Wildlife Society Bulletin* 12:370–376.

Lacy, R. C. 1994. Managing genetic diversity in captive populations of animals. Pages 63–89 in M. L. Bowles and C. J. Whelan (eds.), *Restoration of Endangered Species: Conceptual Issues, Planning, and Implementation.* Cambridge: Cambridge University Press.

Lacy, R. C., J. D. Ballou, F. Princee, A. Starfield, and E. A. Thompson. 1995. Pedigree analysis for population management. Pages 57–75 in J. D. Ballou, M. Gilpin, and T. J. Foose (eds.), *Population Management for Survival and Recovery: Analytical Methods and Strategies in Small Population Conservation.* New York: Columbia University Press.

Lande, R. 1987. Extinction thresholds in demographic models of territorial populations. *American Naturalist* 30(4):624–635.

Lande, R., and G. F. Barrowclough. 1987. Effective population size, genetic variation, and their use in population management. Pages 87–123 in M. E. Soulé (ed.), *Viable Populations.* New York: Cambridge University Press.

Levins, R. 1970. Extinctions. Pages 75–107 in *Some Mathematical Questions in Biology.* Lectures on Mathematics in the Life Sciences. Vol. 2. Providence, R.I.: American Mathematical Society.

Mace, G. M., J. M. Pemberton, and H. F. Stanley. 1992. Conserving genetic diversity with the help of biotechnology—desert antelopes as an example. Pages 123–134 in H.D.M. Moore, W. V. Holt, and G. M. Mace (eds.), *Biotechnology and the Conservation of Genetic Diversity.* Symposia of the Zoological Society of London, No. 64. Oxford: Oxford University Press.

Marsh, D. M., and P. C. Trenham. 2001. Metapopulation dynamics and amphibian conservation. *Conservation Biology* 15:40–49.

Moore, H.D.M., W. V. Holt, and G. M. Mace (eds.). 1992. *Biotechnology and the Conservation of Genetic Diversity.* Symposia of the Zoological Society of London, No. 64. Oxford: Oxford University Press.

Morrison, M. L., C. M. Kuehler, T. A. Scott, A. A. Lieberman, W. T. Everett, R. B. Phillips, C. E. Koehler, P. A. Aigner, C. Winchell, and T. Burr. 1995.

San Clemente loggerhead shrike: Recovery plan for an endangered species. *Proceedings of the Western Foundation of Vertebrate Zoology* 6:293–295.

Morrison, M. L., L. S. Mills, and A. J. Kuenzi. 1996. Study and management of an isolated, rare population: The Fresno kangaroo rat. *Wildlife Society Bulletin* 24:602–606.

Morrison, M. L., B. G. Marcot, and R. W. Mannan. 1998. *Wildlife-Habitat Relationships: Concepts and Applications.* 2nd ed. Madison: University of Wisconsin Press.

Morrison, M. L., L. S. Hall, S. K. Robinson, S. I. Rothstein, D. C. Hahn, and T. D. Rich (eds.). 1999. Research and management of the brown-headed cowbird in western landscapes. *Studies in Avian Biology* 18:204-217.

Nagorsen, D. W., and R. M. Brigham. *1993 Bats of British Columbia.* Vancouver: University of British Columbia Press.

Office of Technology and Assessment (OTA). 1993. *Harmful Non-indigenous Species in the United States.* OTA-F-565. 2 vols. Washington, D.C.: U.S. Congress, Office of Technology Assessment.

Pearce, J., and D. Lindenmayer. 1998. Bioclimatic analysis to enhance reintroduction biology of the endangered helmeted honeyeater (*Lichenostomus melanops cassidix*) in southeastern Australia. *Restoration Ecology* 6:238–243.

Ramey, R. R., II, G. Luikart, and F. J. Singer. 2000. Genetic bottlenecks results from restoration efforts: The case of bighorn sheep in Badlands National Park. *Restoration Ecology* 8:85–90.

Ruggiero, L. F., R. S. Holthausen, B. G. Marcot, K. B. Aubry, J. W. Thomas, and E. C. Meslow. 1988. Ecological dependency: The concept and its implication for research and management. *North American Wildlife and Natural Resources Conference* 53:115–126.

Scott, T. A. 1994. Irruptive dispersal of black-shouldered kites to a coastal island. *Condor* 96:197–200.

Singer, F. J., C. M. Papouchis, and K. K. Symonds. 2000. Translocations as a tool for restoring populations of bighorn sheep. *Restoration Ecology* 8:6–13.

Soulé, M. E. 1980. Thresholds for survival: Maintaining fitness and evolutionary potential. Pages 151–170 in M. E. Soulé and B. A. Wilcox (eds.), *Conservation Biology: An Evolutionary-Ecological Perspective.* Sunderland, Mass.: Sinauer Associates.

———. 1990. The onslaught of alien species, and other challenges in the coming decades. *Conservation Biology* 4:233–239.

Thompson, J. R., V. C. Bleich, S. G. Torres, and G. P. Mulcahy. 2001. Translocation techniques for mountain sheep: Does the method matter? *Southwestern Naturalist* 46:87–93.

Trammell, M. A., and J. L. Butler. 1995. Effects of exotic plants on native ungulate use of habitat. *Journal of Wildlife Management* 59:808–816.

Trulio, L. A. 1995. Passive relocation: A method to preserve burrowing owls on disturbed sites. *Journal of Field Ornithology* 66:99–106.

Tyser, R. W., and C. A. Worley. 1992. Alien flora in grasslands adjacent to road and trail corridors in Glacier National Park, Montana (U.S.A.). *Conservation Biology* 4:251–260.

Wilson, D. E., F. R. Cole, J. D. Nichols, R. Rudran, and M. S. Foster (eds.). 1996. *Measuring and Monitoring Biological Diversity; Standard Methods for Mammals*. Washington, D.C.: Smithsonian.

Wolf, C. M., B. Griffith, C. Reed, and S. A. Temple. 1996. Avian and mammalian translocations: Update and reanalysis of 1987 survey data. *Conservation Biology* 10:1142–1154.

Wright, S. 1931. Evolution in Mendelian populations. *Genetics* 16:97–159.

CHAPTER 2

Habitat

Habitat is considered one of the few unifying concepts in contemporary wildlife ecology. This conclusion is based on numerous studies that relate the presence, abundance, distribution, and diversity of animals to aspects of their environments—studies in which habitat is invoked to explain the evolutionary history and fitness of animals (Block and Brennan 1993). Others have likewise emphasized the importance of wildlife/habitat relationships. "Habitat use" by wildlife has been addressed by numerous researchers. (See the reviews in Verner et al. 1986; Bookhout 1994; and Morrison et al. 1998.) But as Hall et al. (1997) have pointed out, there are several problems with current studies and discussions of habitat use that lead to ambiguity and inaccuracy—a situation that bedevils communication among researchers and confuses land managers and restorationists who are attempting to implement research findings. Restorationists cannot be expected to incorporate the needs of wildlife into a project if the literature is confusing and contradictory.

Although many contend that studies of wildlife/habitat relationships have to be placed in the proper spatial and temporal context (Wiens 1989a; Morrison et al. 1992; Block and Brennan 1993; Litvaitis et al. 1994; Bissonette 1997), this has yet to happen to any great extent. Researchers must recognize that their perceptions of wildlife/habitat relationships depend on the different scales at which different animals operate—and at which we operate (Wiens 1989a; Huxel and Hastings 1999). Johnson (1980) and Hutto (1985), for example, have proposed that animals select habitat through a hierarchical spatial scaling process: selection occurs first at the scale of the geographic range; it occurs second at the scale where animals conduct their activities (that is, in their home ranges); it occurs third at the scale of specific sites or for specific components within their home ranges; and fourth, animals select how they will procure resources within these microsites. Hutto (1985) thinks that selection at the scale of the geographic range is probably genetically determined. Wecker (1964) and Wiens (1972) have demonstrated that selection at finer scales may be influenced by learning and experience and hence under the animal's control. Because wildlife/habitat relationships may be distinctly different at different scales, habitat researchers must be sure to state the scale at which their study is focused and be careful not to extrapolate their data beyond this scale. As Askins (2000) points out, restoration of animals demands an understanding of specific species, which depend on specific types of vegetation, breeding sites, and food.

In terms of temporal scale, researchers must specify when their study occurred and state the time period to which it applies. Morrison et al. (1998:168–172) point out that too many researchers ignore that temporal variation in resource use occurs—or if they do recognize this fact, they still sample only from narrow time periods where the resulting wildlife/habitat relationships apply minimally to other situations. Alternatively, researchers commonly sample from across broad time periods (years or summer or winter seasons) and then use averaged values for variables across the periods—a practice that may mask differences in resource use. Thus a restoration plan must consider the resources each species needs throughout the period of occupancy of the site, whether for a single season (say, winter) or throughout the year. If the species is resident, researchers must determine how its resource needs change with the seasons (for a bird that eats seeds in the winter, for instance, but insects in the summer).

If we want to advance wildlife ecology and thus wildlife restoration, we must be sure that our fundamental concepts are well defined and hence well understood. This not only improves discussion among ecologists by forcing us to use words scientifically and consistently. It also improves our discussions with managers, administrators, and the public, so that our answers are not confusing and ambiguous.

Peters (1991:76) has written about "operationalizing" ecological concepts if environmental scientists hope to further their science. By this, he means that concepts such as habitat should have operational definitions: practical, measurable specifications of the ranges of specific phenomena the terms represent. The definitions may change, of course, but if the concepts are to be scientifically useful, then the definitions must be sufficiently measurable that users can apply them consistently.

Block and Brennan (1993) and Hall et al. (1997) charge that definitions of the term *habitat* are often vague—ranging from how species are associated with broad, landscape-scaled vegetation to detailed descriptions of the immediate physical environment used by species. It is easy to recognize a similar tendency among studies in wildlife science. This vagueness and variability is nonproductive because it detracts from the ability to communicate effectively about habitat and related subjects.

A lack of explicit definition leads ecologists to a variety of approaches for measuring terms such as habitat use, selection, preference, and carrying capacity (Wiens 1984:398), making it extremely difficult for us to conduct comparisons within and between disciplines. Because standard definitions are rarely used, some writers have simply thrown up their hands at trying to provide them (Verner et al. 1986:xi). I think, however, that the prevalence of the word *habitat* in the wildlife, restoration ecology, and conservation biology literature, as well as prevalence of words related to habitat (such as community, ecosystem, and biodiversity), obliges us to develop standard definitions. If restorationists (and other resource managers) are to incorporate new ideas into their plans, it behooves all scientists to ensure their research results are clear and accessible to people from different backgrounds.

When Hall et al. (1997) reviewed papers from prominent journals and books in wildlife and ecology that discussed wildlife/habitat relationships, they examined *habitat* and related terms for use and consistency. Of the 50 articles they reviewed, 47 used the term *habitat*; of these articles, habitat was defined and used correctly (that is, in a species-specific context) in only 5 of the 47 papers (11 percent). The word was used weakly or poorly (without a definition, for example, or sometimes confused with a vegetation association) in 34 of the 47 papers (72 percent). It was used incorrectly (not defined, for example, and always confused with a vegetation association) in 8 of the 47 papers (17 percent). The term most often used incorrectly was *habitat type*; in only one instance was the word used as first defined by Daubenmire (1968).

Definitions

The definitions presented here are based on Block and Brennan (1993), Hall et al. (1997), and Morrison et al. (1998), who in turn based them on the original intent of such ecologists as Grinnell (1917), Leopold (1933), Hutchinson (1957), Daubenmire (1968), and Odum (1971). In addition to habitat—the focus of this chapter—I discuss the terms *niche*, *landscape*, and *resources*. I consider these definitions here because of their frequent use in wildlife and restoration ecology.

Habitat

I define *habitat* as the resources and conditions present in an area that affect occupancy by a species. Habitat is organism-specific: it relates the presence of a species, population, or individual (animal or plant) to an area's physical and biological characteristics. Habitat involves more than vegetation or vegetation structure; it is the sum of the specific resources needed by a species. Wherever an organism is provided with resources that affect its ability to survive, that is habitat. Migration corridors, dispersal corridors, and the land that animals occupy during breeding and nonbreeding seasons—all are habitat. Thus, habitat is not equivalent to *habitat type*, a term coined by Daubenmire (1968:27–32) that refers only to the type of vegetation association in an area or the potential of vegetation to reach a specified climax stage. Habitat is much more than an area's vegetation (such as pine-oak woodland). The term *habitat type* should not be used when discussing wildlife/habitat relationships. When we want to refer only to the vegetation that an animal uses, we should say *vegetation association* or *vegetation type* instead.

The confusion between habitat and habitat type has led to a general misconception about how to restore an area for wildlife. If habitat is species-specific, then any plot of land has numerous habitats; each habitat corresponds to a specific species. As you gaze across an area, therefore, you are viewing numerous habitats of likely different quality. Thus the definition of habitat as species-specific is an absolutely critical concept. It means that restoring vegetation, regardless of how well it matches some desired condition, can easily fail to restore the desired assemblage of wildlife. Failure to plan simultaneously for plant and animal restoration results in a hit-or-miss strategy for animals and clearly falls under the Field of Dreams hypothesis—"if you build it, they will come" (Palmer et al. 1997:295). Restoring vegetation restores wildlife habitat for *some* species, but not necessarily the species desired. Poor planning for wildlife may create an ecological trap in which an undesired species kills or harasses a desired species or its young.

I define the term *habitat use* as the way an animal uses (or "consumes," in a generic sense) a collection of physical and biological components (that is, resources) in a habitat. With respect to habitat selection, as mentioned previously, Hutto (1985:458) proposed that it is a hierarchical *process* involving a series of innate and learned behavioral decisions made by an animal about what habitat it will use at different scales of the environment. Likewise, Johnson (1980) refers to selection as the process by which an animal chooses which habitat components to use. Given the body of literature supporting the view of selection as a process, it is useful to define selection this way and hence to define *habitat preference* as the consequence of the process, resulting in the disproportional use of some resources over others.

Habitat availability refers to the accessibility of physical and biological components needed by animals—as opposed to the abundance of these resources, which refers only to their quantity in the habitat irrespective of the organisms present in the habitat (Wiens 1984:402). In theory, we should be able to measure the amounts and kinds of resources available to animals; in practice, however, it is often impossible to assess resource availability from an animal's point of view (Litvaitis et al. 1994). We can measure the abundance (by trapping) of a prey species for a particular predator, for example, but we cannot say that all of the prey present in the habitat are available to the predator because there may be constraints, such as presence of ample cover, that restrict their accessibility. Similarly, vegetation beyond the reach of an animal is unavailable for it to feed on, even though the vegetation may be its preferred forage. Although measuring actual resource availability is important for understanding wildlife/habitat relationships, in practice it is seldom measured because of the difficulty in determining exactly what is available and what is not (Wiens 1984:406). Consequently, quantifying availability usually consists of a priori or a posteriori measures of the abundance of resources in an area used by an animal, rather than actual availability. Thus in most instances the term *availability* should be avoided by biologists. The term *abundance* should be used instead because this is what is most commonly measured. Where the accessibility of a resource has in fact been determined for an animal, analyses to assess habitat preference by comparing use versus availability are valuable.

The term *habitat quality* refers to the ability of the environment to provide conditions appropriate for individual and population persistence. Quality is a continuous variable ranging from low- to medium- to high-quality habitats based on their ability to provide resources for survival, reproduction, and population persistence. Researchers commonly equate high-quality habitat with vegetative features that may contribute to the presence (or absence) of a species (as in Habitat Suitability Index models; see Laymon and Barrett 1986 and Morrison et al. 1991). Quality must be explicitly linked with demographic features, however, if it is to be a useful measure. Discussions of carrying capacity (Leopold 1933; Dasmann et al. 1973), for example, have equated a high-quality habitat with one that has a density of animals in balance with its resources. In the field, this often means giving a high rank to habitats with large densities of animals (Laymon and Barrett 1986). Van Horne (1983) has demonstrated that density is a misleading indicator of habitat quality, however, and the widespread occurrence of source and sink habitats in nature (Pulliam 1988; Wootton and Bell 1992) has persuaded many ecologists to deemphasize this ranking. Thus while carrying capacity may be equated with a certain level of habitat quality, the quality itself should be based not on the number of organisms but on the demographics of individual populations.

For a restorationist, the key concept is habitat quality. If your project's goal is to restore a viable population of breeding individuals, for example, the critical factors causing the species to survive and reproduce successfully must be present. And as we have seen, these factors go far beyond vegetation and include food (say, the specific arthropods in the vegetation), breeding sites of proper condition (say, shaded nest sites), and perhaps an absence of predators.

Terms such as *macrohabitat* and *microhabitat* are relative and refer to the scale at which a study is being conducted for the animal in question (Johnson 1980). Thus macrohabitat and microhabitat must be defined for each study on a species-specific basis. Generally, macrohabitat refers to broad-scaled features such as seral stages or zones of specific vegetation associations (Block and Brennan 1993)—which usually equate to Johnson's first level of habitat selection. Microhabitat usually refers to fine-scaled habitat features—which are important factors at levels 2 to 4 in Johnson's hierarchy. Thus it is appropriate to use the terms *microhabitat* and *macrohabitat* in a relative sense, and the scales to which they apply should be stated explicitly.

It should be evident, then, that quantifying habitat use can be very complicated—and this makes it hard to predict a species' distribution and its ability to colonize restored sites. Determination of key habitat factors, however, will identify the conditions where the species might occur. Such information leads to restoration actions (such as plant species composition and structure) and management actions (such as control of predators or competitors) that could allow for occupancy of a site not being used by the species.

Niche

Wiens (1989a:146) has called the *niche* one of the most variably defined terms in ecology. Two primary meanings have been given to the term. A species' *Grinnellian niche* is the range of environmental features that enable individuals to survive and reproduce. Grinnell's (1917) focus was on factors determining the distribution and abundance of species. The *Eltonian niche*, in contrast, describes the niche of a species as its functional role in the community, especially with regard to trophic interactions (Elton 1927). Hutchinson (1957) expanded this concept of the niche by mathematically describing a large number of environmental dimensions, each representing some resource or other important factor on which different species exhibit frequency distributions of performance, response, or resource utilization (Wiens 1989a:146). Each perspective results in a different emphasis of study: studies of individuals under the Grinnellian view and studies of communities under the Eltonian-Hutchinson view.

Arthur (1987) recommends that we follow MacArthur's (1968) quantification of the niche, which plots utilization against some quantifiable resource variable that he calls the *resource utilization function* (RUF). Arthur thinks it is better to build complexity as needed, as with RUFs, than to dissect it using some multidimensional concept. RUFs describe the choice of resources by animals; these choices may be constrained by predators, competitors, and other factors. I prefer this approach because it makes far fewer assumptions about organizational structure and can be tuned to fit specific questions.

Thus the habitat contains the resources that affect occupancy, survival, and reproduction, whereas the niche concerns access to these resources and use of them. These concepts raise critical issues in restoration planning, for we see that simply providing the resources might be inadequate to ensure that the restoration goal is met. For example, a restorationist must be concerned with the distribution and abundance of competitors for resources that are being planned for a target species. It does little good to provide food if the animal of interest will get killed trying to harvest it. Thus restoration might entail removing certain features that allow the competitor to occupy the site or even removing the competitor itself.

Landscape

Landscape can be defined as a spatially heterogeneous area used to describe features of interest (stand type, site, soil). King (1997:205–206) describes a landscape primarily by its spatial extent. A serious problem with application of the term *landscape* is that it is usually taken to mean a large area (1–100 km²) (Forman and Gordon 1986; Davis and Stoms 1996). The perception of "landscape" to a small animal, however, is likely much different than that perceived by a large one. As King (1997:204) has noted, the fundamental themes of landscape ecology do not just apply to areas greater than a few square kilometers. The influence of spatial heterogeneity on biotic and abiotic processes can be addressed at virtually any spatial scale. Thus we should not place area limitations on the notion of landscape. Although describing landscape in terms of square kilometers is appropriate for certain applications (such as placing a restoration project in the context of a broad area), describing it in terms of a few square meters is appropriate for other uses (such as salamander/niche relationships).

Resources

Wiens (1989b:262) notes that although resources are involved in most explanations of community patterns, all too often they have been defined in ad hoc ways. Rarely are they measured directly or inferred to be limiting without any evidence. Thus to define a resource, the area of interest must be explicitly identified with respect to its spatial extent and broken down into its measurable elements.

Little attention has been given the identification and measurement of resources. As Wiens (1989a:321) has indicated, almost any environmental factor that correlates with the distribution, abundance, or reproductive performance of a species has been called a resource. But without a precise definition of the resources present, it is impossible to derive accurate patterns of resource use or niche relationships. I define a *resource* as any biotic or abiotic factor that is directly used by an organism. Resources that are limiting to an organism could then be referred to as *limiting resources.* Wiens (1989a:321–323) has also noted that the differences between resource abundance, availability, and use must be distinguished to be certain which one is actually being measured. *Resource abundance* is the item's absolute amount (or size or volume) in an explicitly defined area—for example, the number of food items in 1 ha. *Resource availability* is the amount of a resource actually available to the animal (that is, the amount exploitable)—for example, the number of food items in 1 ha that an ungulate can reach. *Resource use* is a measure of the amount of the resource directly taken (consumed, removed) from an explicitly defined area—for example, the number of food items in 1 ha that an animal consumes in a six-hour sampling period.

Determining the critical resources necessary for a target species—and identifying any constraints on the use of these resources—is a fundamental aspect of wildlife restoration. Here again, there are numerous natural history papers and general field guides that provide at least rudimentary information on these factors. The restorationist can list the species of interest to restoration and then identify the key resources—and constraints on their use—for each species. Many similarities among species will likely be evident. You can then use this list in your planning to maximize the opportunity for the species to actually use the restored site.

When to Measure

The behavior, location, and needs of animals change, often substantially, throughout the year. Many researchers ignore temporal variations in habitat use, however, which can undermine habitat assessments. Without knowledge of an animal's total requirements, restoration plans have limited and perhaps faulty implications.

The decision on when to measure is a study-specific problem determined by the natural history of the species of interest. Species that are permanent residents in the project area should be studied throughout the year. More and more studies are showing that animals often change their use of resources substantially between seasons (Schooley 1994; Morrison et al. 1998:168–172). The arthropod fauna available to birds, for example, shifts between species of trees as the seasons' change. Failure to provide the proper mixture of plant species could easily result in failure of a restoration project regardless of the vigor of the plants that are established. Intuitively we would expect that the fall and winter periods—when populations are at their greatest numbers (because of offspring), resources are declining (trees and arthropods are dying or going dormant), animals are physiologically stressed by dispersal or migration, and the weather is becoming more harsh—are the most difficult times for an animal.

What to Measure

Green (1979:10) has listed several criteria that should be considered when you are choosing variables to measure wildlife habitat:

- Spatial and temporal variability in biotic and environmental variables used to describe or predict impacts
- Feasibility of sampling with precision at a reasonable cost
- Relevance to the impacts and sensitivity of response to them

These criteria apply both in descriptive studies and in analyses of impacts (chemical spills, forest harvesting). Understanding the variability in the system of interest is critical in designing a restoration project. This variability includes natural, stochastic, or systematic change as well as measurement and sampling error. Researchers conduct a cost/benefit analysis either formally or informally when choosing variables for measurement: they must determine the precision necessary to reach the project's goals and then match all the sampling to this level of precision. Is the goal of the project to provide for simple presence or absence of a species, a specific density, or reproduction? These are questions the restorationist must answer before designing a project.

Spatial Scale

You must match the scale of analysis with the scale you wish to apply in restoration and management. In restoration, these decisions are driven by the size of the project area and the goals regarding wildlife. In general, the smaller the area, the more attention you must give to microhabitat parameters of the species of interest. This is because the probability that a specific habitat component will occur naturally increases as area size increases. For example, a large area is more likely than a small one to contain snags (standing dead trees), a rock outcrop, a pond, or a woodland stand.

The definition of "small" and "large" depends on the project. Many salamanders have home ranges of under 15 m^2 and are unlikely to move over 25 m (Grover 1998). Bratton and Meier (1998) note that salamanders in the southern Appalachians must be considered carefully during plant restoration: because salamanders move only short distances and will not cross even narrow dry areas, the scale of restoring salamanders is finer than that for most plants. The home ranges of other small to medium-sized terrestrial vertebrates, by contrast, can be 5 to 10 ha or more. Projects focused on one or just a few species must be guided by the natural history of these animals. Projects of larger scale that have more general goals (such as enhancing vertebrate diversity) must be guided by principles that relate general measures of the wildlife community (say, species richness) to general measures of the environment (say, vegetation structure).

The finer-scale (microhabitat) relationships almost always vary between locations and time periods and certainly between populations. The magnitude of these variations determines the generality of the model. (Generality refers to the model's applicability at other times and places.) Much of the wildlife/habitat literature has been criticized because of its time and place specificity (Irwin and Cook 1985). This criticism is misplaced, however, and shows a general lack of understanding of the relationship between the precision of the variables measured and the scale of application possible. The decision to develop broad-scale models versus fine-scale models should be based on the objectives of the study. The extensive approach cannot tell us how an animal reacts to changes in litter depth, or the local density of trees by species, or the occurrence of a predator in a specific patch of vegetation. Such details are necessary, however, for management of local populations of animals.

Wildlife managers get frustrated when models fail to work in their specific location. This frustration comes primarily from trying to apply a relationship based on broad measurements of vegetation to local situations. Likewise, models developed at a fine scale can seldom be generalized to other locations (Block et al. 1994). For the restorationist, this means you must give careful consideration to matching the type of information available to the specific size and characteristics of the project area.

Measurements: Conceptual Framework

Two basic aspects of vegetation must be distinguished: its structure (physiognomy) and the taxa of the plants (floristics). (See Figure 2.1.) Many ecologists initially concluded that vegetation structure and "habitat configuration" (size, shape, and distribution of vegetation in an area)—rather than plant taxonomic composition—were paramount in determining patterns of habitat occupancy by animals, especially birds. (See the review by Morrison et al. 1998:146–147.) But recent studies have shown that plant species com-

FIGURE 2.1.
The height and layering of vegetation, as well as the species composition of the plants, play central roles in determining an animal's use of habitat. Depicted here is remnant riparian vegetation along the lower Colorado River, Arizona. (Photo courtesy of Annalaura Averill-Murray and Suellen Lynn.)

position plays a greater role in determining patterns of habitat occupancy than previously thought. The relative usefulness of structural versus floristic measures is again primarily a function of the spatial scale of analysis.

A species that appears to respond to the physical configuration of the environment (its physiognomy) at the continental level may show little correlation with physiognomy at the regional or local level. Thus many animals may differentiate between gross vegetative types on the basis of physiognomy, with further refinement of the distribution (and thus abundance) within a local area based on plant taxonomic considerations.

MACROHABITAT AND MICROHABITAT. With the rise in studies of animal diversity, researchers began to develop various measures to relate the numbers and kinds of animals to the gross structure of the vegetation. Most famous is the relationship between foliage height diversity (FHD) and bird species diversity (BSD): as foliage layers are added, the number of bird species tends to increase (see Figure 2.2). In vertically simple vegetation, such as brushland and grassland, FHD would not be expected to provide a good indicator of animal diversity (at least for most vertebrates). Recognizing this problem, Roth (1976) developed a method by which the dispersion of clumps of vegetation such as shrubs forms the basis for a measure of habitat heterogeneity or "patchiness." In fact, Roth was able to relate BSD to this patchiness.

Returning to Figure 2.2, note that there is considerable scatter around the regression line. Thus the usefulness of this general principle as a site-specific predictor declines as the scale of application becomes increasingly fine—that is, as you go from macroscale to microscale, or from what are usually termed "landscape" projects to local projects. Measures of diversity sacrifice complexity for simplicity; this is why they are useful primarily at larger spatial scales. These indices collapse detailed information on plants—such as species composition, foliage condition (vigor), and arthropod abundance—into a single number.

Many of the current habitat models operate at the macrohabitat scale, including most statewide constructs of wildlife/habitat relationships (WHR) (Block et al. 1994), GAP models (Scott et al. 1993), and habitat suitability index (HSI) models (USFWS 1981). Most of these models use broad-scale categorizations of vegetation types (often mislabeled as "habitat types") as a predictor of animal presence or abundance. But many developers of these models substantially mismatch scales in the variables used to develop their models; this is especially evident in HSI and WHR models. Such mismatch-

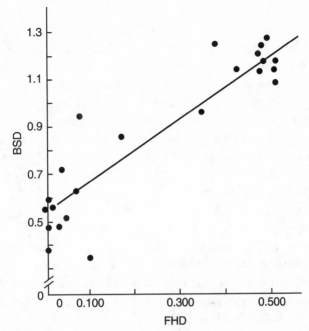

FIGURE 2.2.
Foliage height diversity (FHD) versus bird species diversity (BSD). Solid dots represent the study sites. (From M. F. Willson, "Avian Community Organization and Habitat Structure," Figure 1. *Ecology* 55:1017–1029. Copyright 1974.)

ing (entering microhabitat and macrohabitat variables into the same analysis, for example) ignores current theories concerning the hierarchic nature of habitat selection and makes it difficult to interpret a model's output. Models developed at the macrohabitat level help us to understand broad habitat relationships, but they should be limited to application at the broad scale.

Thus restoration projects that occur on relatively small areas (less than several km²) usually concentrate on microhabitat relationships. And when the goal is to obtain successful survival and reproduction on the area, microhabitat factors and niche relationships (such as constraints on resource use) become the focus. Projects seeking to enhance biodiversity across large areas, in contrast, are likely to concentrate on gross measures of vegetation structure.

THE FOCAL-ANIMAL APPROACH. Most studies of microhabitat selection are variations of the *focal-animal approach*. These methods use the presence of an animal as an indication of the habitat being used by the species. No correlation between abundance and the environment is involved. Rather, the location of individual animals is used to demark an area from which environmental variables are measured. As detailed in the following section, an animal's specific location might serve as the center of a sampling plot. Or a series of observations of an individual might be used to delineate an area from which samples are then made. (See, for example, Wenny et al. 1993.) In either case, the major assumption is that measurements indicate the animal's habitat preferences. Many studies, for example, have used the location of a singing male bird or a foraging individual as the center of plots describing the habitat of the species (James 1971; Holmes 1981; Morrison 1984a, 1984b; VanderWerf 1993).

EXAMPLES. To what specific aspect of vegetation are animals responding? What are the stimuli causing the behavior that we call resource use? To answer these questions, we will consider a few examples of variables collected by researchers seeking to describe the habitat-use patterns of animals. This section is meant to give you a sense of the types of data you will need to design projects for the specific species.

James (1971) conducted one of the first and most-cited studies quantifying bird/habitat relationships. Using 15 measures of vegetation structure to describe the multidimensional "habitat space" of a bird community in Arkansas, she followed closely the methods that she and a colleague had developed (James and Shugart 1970). These methods are described in the next section. The conceptual framework and general analytic techniques (multivariate analysis of focal-bird observations) have led to a plethora of studies expanding upon her basic ideas. Indeed her strategy and methods are still in wide use. Murray and Stauffer (1995), for example, based their vegetation sampling on the James and Shugart methodology.

Dueser and Shugart (1978) had as their goal the description of microhabitat differences among the small-mammal species of an upland forest in eastern Tennessee. Their specific objectives were to characterize and compare microhabitats of species within the forest and to examine how species abundance and distribution relate to the availability of selected microhabitats. They gathered information for vertical strata at each capture site of a small mammal: overstory, understory, shrub level, forest floor, and litter-soil level. Table 2.1 lists the variables they collected. Note that they did not collect species-specific information on plants beyond designations of "woodiness," "evergreenness," and the like—an unfortunate omission for a microhabitat analysis, the ramifications of which are unknown. They paid special atten-

TABLE 2.1. Sampling Methods for Variables Measuring Forest Habitat Structure

Variable	Sampling method
1. Percentage of canopy closure	Percentage of points with overstory vegetation—from 21 vertical ocular tube sightings along the center lines of two perpendicular 20-m^2 transects centered on trap
2. Thickness of woody vegetation	Average number of shoulder-height contacts (trees and shrubs)—from two perpendicular 20-m^2 transects centered on trap
3. Shrub cover	Same as variable (1)—for presence of shrub-level vegetation
4. Overstory tree size	Average diameter (in cm) of nearest overstory tree—in quarters around trap
5. Overstory tree dispersion	Average distance (m) from trap to nearest understory tree—in quarters
6. Understory tree size	Average diameter (cm) of nearest understory tree—in quarters around trap
7. Understory tree dispersion	Average distance (m) from trap to nearest understory tree—in quarters
8. Woody stem density	Live woody stem count at ground level within a 1-m^2 ring centered on trap
9. Short woody stem density	Live woody stem count within a 1-m^2 ring centered on trap (stems $\leq$ 0.40 m in height)
10. Woody foliage profile density	Average number of live woody stem contacts with an 0.80-cm-diameter metal rod rotated 360° describing a 1-m^2 ring centered on the trap and parallel to the ground at heights of 0.05, 0.10, 0.20, 0.40, 0.60, . . . , 2 m above ground level
11. Number of woody species	Woody species count within a 1-m^2 ring centered on trap
12. Herbaceous stem density	Live herbaceous stem count at ground level within a 1-m^2 ring centered on trap
13. Short herbaceous stem density	Live herbaceous stem count within a 1-m^2 ring centered on trap (stems $\leq$ 0.40 m in height)
14. Herbaceous foliage profile density	Same as variable (10)—for live herbaceous stem contacts
15. Number of herbaceous species	Herbaceous species count within a 1-m^2 ring centered on trap
16. Evergreenness of overstory	Same as variable (1)—for presence of evergreen canopy vegetation
17. Evergreenness of shrubs	Same as variable (1)—for presence of evergreen shrub-level vegetation
18. Evergreenness of herb stratum	Percentage of points with evergreen herbaceous vegetation—from 21 step-point samples along the center lines of two perpendicular 20-m^2 transects centered on trap

(continues)

TABLE 2.1. *Continued*

Variable	Sampling method
19. Tree stump density	Average number of tree stumps ≥ 7.50 cm in diameter—per quarter
20. Tree stump size	Average diameter (cm) of nearest tree stump ≥ 7.50 cm in diameter—in quarters around trap
21. Tree stump dispersion	Average distance (m) to nearest tree stump ≥ 7.50 cm in diameter—in quarters around trap
22. Fallen log density	Average number of fallen logs ≥ 7.50 cm in diameter—per quarter
23. Fallen log size	Average diameter (cm) of nearest fallen log ≥ 7.50 cm in diameter—in quarters around trap
24. Fallen log dispersion	Average distance (m) from trap to nearest fallen log ≥ 7.50 cm in diameter—in quarters around trap
25. Fallen log abundance	Average total length (> 0.50 m) of fallen logs ≥ 7.50 cm in diameter—per quarter
26. Litter-soil depth	Depth of penetration (< 10 cm) into litter-soil material of a hand-held core sampler with 2-cm-diameter barrel
27. Litter-soil compactibility	Percentage of compaction of litter-soil core sample (variable 26)
28. Litter-soil density	Dry weight density (g/cm^2) of litter-soil core sample (variable 26) after oven drying at 45°C for 48 hr
29. Soil surface exposure	Same as variable (18)—for percentage of points with bare soil or rock

Source: R. D. Dueser and H. H. Shugart, Appendix. *Ecology* 59:89–98. Copyright 1978. Reproduced by permission of the Ecological Society of America.

tion to features of the forest floor—such as litter-soil compactability, fallen log density, and short herbaceous stem density—and found that certain of these soil variables played a significant role in describing the differences in microhabitats of the species studied.

Morrison et al. (1995) used time-constrained surveys to describe the microhabitats of amphibians and reptiles in the mountains of southeastern Arizona. Observers walked slowly, searching the ground and tree trunks and turning over movable rocks, logs, and litter to examine protected locations while a stopwatch ran. When the survey time stopped, a 5-m-diameter plot was then centered on the animals' location and served as the site where microhabitat conditions were measured: substrate temperature, various aspects of the vegetation, and other habitat characteristics.

Welsh and Lind (1995) analyzed the habitat affinities of the Del Norte salamander (*Plethodon elongates*) in relation to landscape, macrohabitat, and microhabitat scales. They presented a detailed rationale for the selection of methods, including choice of analytic techniques, data screening, and interpretation of output. The variables they measured, by spatial scale, are shown in Table 2.2.

These examples represent a useful starting point for designing a study of wildlife/habitat relationships. A note of caution: Do not try to duplicate the methods used in these studies exactly. Rather, select the variables that appear to predict something of interest about the species being studied. Gathering

TABLE 2.2. Hierarchic Arrangement of Ecological Components Represented by 43 Measurements of the Forest Environment Taken in Conjunction with Sampling for the Del Norte Salamander (*Plethodon elongates*)

HIERARCHIC SCALE
  Variable category
    Variables[a]

I. BIOGEOGRAPHIC SCALE[b]

II. Landscape scale
  A. Geographic relationships
    • Latitude (degrees)
    • Longitude (degrees)
    • Elevation (m)
    • Slope (%)
    • Aspect (degrees)

III. MACROHABITAT OR STAND SCALE
  A. Trees: density by size[c]
    • Small conifers (C)
    • Small hardwoods (C)
    • Large conifers (C)
    • Large hardwoods (C)
    • Forest age (in years)
  B. DEAD AND DOWN WOOD: SURFACE AREA AND COUNTS
    • Stumps (B)
    • All logs—decayed (C)
    • Small logs—sounds (C)
    • Sound log area (L)
    • Conifer log-decay area
    • Hardwood log-decay area (L)

(continues)

TABLE 2.2. *Continued*

C. SHRUB AND UNDERSTORY COMPOSITION (> 0.5 M)
- Understory conifer (L)
- Understory hardwoods (L)
- Large shrub (L)
- Small shrub (L)
- Bole (L)
- Height II—ground vegetation (B) (0–0.5 m)

D. GROUND-LEVEL VEGETATION (< 0.5 M)
- Fern (L)
- Herb (L)
- Grass (B)
- Height I—ground vegetation (B) (0–0.5 m)

E. Ground cover
- Moss (L)
- Lichen (B)
- Leaf (B)
- Exposed soil (B)
- Litter depth (cm)
- Dominant rock (B)
- Codominant rock (B)

F. Forest climate
- Air temperature (°C)
- Soil temperature (°C)
- Solar index
- % canopy closed
- Soil pH
- Soil relative humidity
- Relative humidity (%)

IV. MICROHABITAT SCALE
A. Substrate composition
- Pebble (P) (% of 32–64 mm diameter rock)
- Cobble (P) (% of 64–256 mm diameter rock)
- Cemented (P) (% of rock cover embedded in soil/litter matrix)

Source: H. H. Welsh and A. J. Lind, "Habitat Correlates of Del Norte Salamander, *Plethodon elongates,* in Northwestern California," Table 1. *Journal of Herpetology* 29:198–210. Copyright 1995. Reproduced by permission of the Department of Zoology, Ohio University

[a] Abbreviations used for the variables: C = count variables (number per hectare); B = Braun-Blanquet variables (percentage of cover in 0.10-ha circle); L = line transect variables (the percentage of 50-m line transects); P = percentage within 49-m^2 salamander search area.

[b] Level I relationships (the biogeographic scale) were not analyzed because all sampling occurred within the range. Spatial scales are arranged here in descending order from coarse to fine resolution.

[c] Small trees = 12–53 cm DBH (diameter at breast height); large trees = > 53 cm DBH.

data is time consuming, and careful planning during the process of variable selection will help you to focus the study.

How to Measure

In this section I review some common methods used to measure wildlife habitat. I cannot survey all the literature available for all taxa here; for a thorough review of basic sampling techniques for all the major taxa of wildlife see Cooperrider et al. (1986) and Bookhout (1994). My intent here is to enhance the restorationist's ability to gather and interpret the information necessary to guide projects aimed at specific wildlife species.

Sampling Principles

As we have seen, vegetation forms the traditional template for how we view wildlife/habitat selection. A cursory review of the methods in wildlife publications shows a reliance on standard techniques of quantifying the structure and floristics of vegetation: point quarter, circular plots and nested circular plots, sampling squares, line intercepts, and so on. These methods are used because they have been tested by plant ecologists in a multitude of environmental situations. Standard methods provide an established starting point from which biologists can adapt specific methods as needed. Standard methods also provide comparability between studies. There are many fine books that review sampling methods in vegetation ecology (Daubenmire 1968; Mueller-Dombois and Ellenberg 1974; Greig-Smith 1983; Cook and Stubbendieck 1986; Bonham 1989; Schreuder et al. 1993).

Sampling Methods

The most popular methods of measuring microhabitat originated with a protocol developed by James and Shugart (1970), who developed a quantitative method of obtaining vegetation data in a simple and standardized manner. Their original intent was to discover a method that could augment the data on bird populations being gathered in the National Audubon Society's "Breeding-Bird Censuses" and "Winter Bird-Population Studies" throughout the United States. But as noted earlier, their strategy has found wide applicability throughout the ecological community. Essentially they gathered data on the density, basal area, and frequency of trees as well as canopy height, shrub density, percent ground cover, and percent canopy cover. They established 0.1-acre (0.04-ha) plots to estimate tree density and frequency. To estimate shrub density they made two transects at right angles to one another across the 0.1-acre plots, counting the number of woody stems intercepted by their outstretched arms. An ocular (sighting) tube was used to estimate vegetation cover. They also provided details on how the sampling equipment could be constructed and offered examples of data sheets.

Earlier we discussed the importance of James's (1971) paper to our conceptualization of how animals perceive their environment. The methods used by James have had a pronounced influence on analyses of wildlife habitat. Circular plots are easy to establish, mark, measure, and relocate, and estimates of animal numbers within such plots can be statistically related to vegetation data in a straightforward manner. Plots provide for the sampling of vegetation and animals at specific locations in space and time. Thus plots are easy to pinpoint using geographic positioning systems (GPS), and their data can then be entered into geographic information systems (GIS). If plots can be considered independent data points (a function of the sampling design and behavior of the animals), then your sample size is equal to the number of plots you sampled. Or if you use the plots to sample from a single study area, they can be averaged and you can calculate associated measures of variance. Noon (1981) has presented a useful description of both the transect and the areal plot sampling systems. The problem with transects is that they cover relatively large areas and thus make it hard to relate specific animal observations (or abundances) to specific sections of the transect. Nevertheless, transects are widely used to provide an overall description of the vegetation of entire study areas.

In sum, then, fixed-area plots and transects can be used for site-specific, detailed analysis of wildlife/habitat relationships. The majority of sampling methods used since the 1970s to develop wildlife/habitat relationships—for subsequent multivariate analyses—have used fixed-area plots (usually circular) as the basis for developing of a sampling scheme that may then incorporate subplots, sampling squares, and transects. Now let us consider some of the more widely used methods.

Dueser and Shugart (1978) developed a detailed sampling scheme that combined plots of various sizes and shapes, as well as short transects (see Figure 2.3). Although designed for analysis of small-mammal habitat, the techniques can easily be adapted for most terrestrial vertebrates. Dueser and Shugart established three independent sampling units centered on each trap: a 1-m^2 ring; two perpendicular 20-m^2 arm-length transects; and a 10-m-radius circular plot. The 1-m^2 circular plot provided a measure of vertical

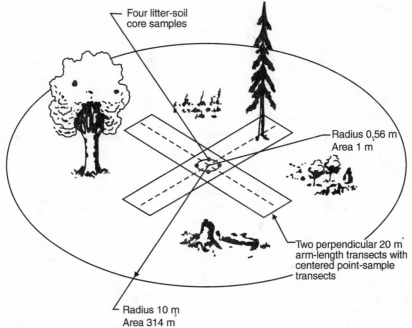

Four litter-soil
core samples

Radius 0.56 m
Area 1 m

Two perpendicular 20 m
arm-length transects with
centered point-sample
transects

Radius 10 m
Area 314 m

FIGURE 2.3.

Habitat variable sampling configuration used by Dueser and Shugart in their study of small-mammal habitat use. (From R. D. Dueser and H. H. Shugart, "Microhabitats in Forest-Floor Small Mammal Fauna," Figure 1. *Ecology* 59:89–98. Copyright 1978.)

foliage profile from the ground through 2-m height for both herbaceous and woody vegetation. Four replicate core-sample estimates of litter-soil depth, compactability, and dry weight density were made on the perimeter of this central ring. The two arm-length transects provided measures of cover type, surface characteristics, and density and evergreenness of the four strata of vegetation. Data recorded for each quarter of the 10-m-radius plot included the species, diameter at breast height, distance from the trap to the nearest understory and overstory trees, numbers of stumps and fallen logs, basal diameter and distance of nearest stump and fallen log, and total length of fallen logs.

In his analysis of snake populations, Reinert (1984) adopted techniques similar to those used in the bird study by James (1971) and the small-mammal study by Dueser and Shugart (1978). In fact, Reinert applied the basic conceptual framework used by the earlier authors in developing the rationale for his methods. Reinert made several modifications of their sampling methods, however. Notably he used a 35-mm camera equipped with a

28-mm wide-angle lens to photograph 1-m² plots from directly above the location of a snake. He then determined the various surface cover percentages by superimposing each slide onto a 10 × 10 square grid. Reinert, then, quantified his measure of cover values more rigorously than most workers, who usually use ocular estimates. His sampling scheme is summarized in Figure 2.4; his variable list is presented in Table 2.3. Note the similarity between Reinert's design and that of Dueser and Shugart, including the minor mixing of spatial scales. Reinert added several environmental variables that measured air, surface, and soil temperature and humidity. The values of these variables obviously depend on the time of day and the general weather conditions at the time of measurement; such constraints do not influence (are not correlated with) the other variables measured. Several popular techniques for quantifying foliage cover are presented in Figure 2.5.

Bibby et al. (1992) have compiled a basic but useful summary of habitat assessment techniques, including a description of mapping techniques for studies of avian ecology and an explanation of how to relate bird counts to environmental characteristics. (See also Chapter 6.) Figure 2.6 shows how

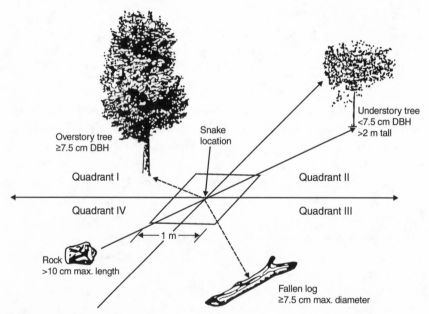

FIGURE 2.4.
Sampling arrangement for snake locations. (DBH = diameter at breast height) (From H. K. Reinert, "Habitat Separation between Sympatric Snake Populations," Figure 1. *Ecology* 65:478–486. Copyright 1984.)

TABLE 2.3. Structural and Climatic Variables Used by Reinert

Mnemonic	Variable	Sampling method
ROCK	Rock cover	Coverage (%) within 1-m² quadrant centered on snake location
LEAF	Leaf litter cover	Same as ROCK
VEG	Vegetation cover	Same as ROCK
LOG	Fallen log cover	Same as ROCK
WSD	Woody stem density	Total number of woody stems within 1-m² quadrant
WSH	Woody stem height	Height (cm) of tallest woody stem within 1-m² quadrant
MDR	Distance to rocks	Mean distance (m) to nearest rocks (>10 cm max. length) in each quarter
MLR	Length of rocks	Mean max. length (cm) of rocks used to calculate MDR
DNL	Distance to log	Distance (m) to nearest log (≥ 7.5 cm max. diameter)
DINL	Diameter of log	Max. diameter (cm) of nearest log
DNOV	Distance to overstory tree	Distance (m) to nearest tree (≥ 7.5 cm DBH)[a]
DBHOV	DBH of overstory tree	Mean DBH (cm) of nearest overstory tree within each quarter
DNUN	Distance to understory tree	Same as DNOV (trees < 7.5 cm, DBH > 2-m height)
CAN	Canopy closure	Canopy closure (%) within 45(cone with ocular tube)
SOILT	Soil temperature	Temperature (°C) at 5-cm depth within 10 cm of snake
SURFT	Surface temperature	Temperature (°C) of substrate within 10 cm of snake
IMT	Ambient temperature	Temperature (°C) of air 1 m above snake
SURFRH	Surface relative humidity	Relative humidity (%) at substrate within 10 cm of snake
IMRH	Ambient relative humidity	Relative humidity (%) 1 m above snake

Source: H. K. Reinert, "Habitat Restoration between Sympatric Snake Populations," Table 1. *Ecology* 65:478–486. Copyright 1984. Reproduced by permission of the Ecological Society of America.

[a] DBH = diameter at breast height.

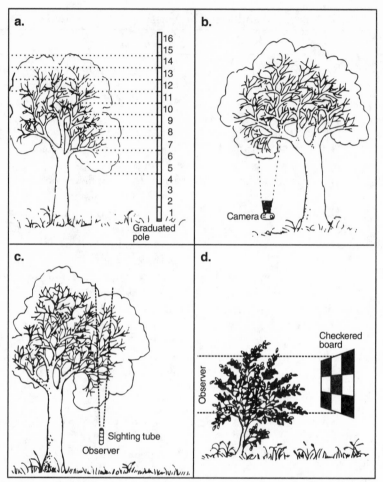

FIGURE 2.5.

Some commonly used devices to measure habitat variables in woodlands. (a) Graduated pole held upright—most useful to measure the features of the foliage in the shrub layer and low forests. (b) 35-mm camera with 135-mm or zoom lens—can be focused down through the forest profile (heights read off rangefinder) and used to assess foliage density through a vertical section of forest. (c) Sighting tube—observer looks directly up and assesses the canopy or shrub layer foliage density or divides the profile into height bands and assesses vegetation cover within each band. (d) Checkered board— used to assess vertical density of shrub layer. Observer walks away from the board until 50 percent of it is judged to be obscured by vegetation; this produces an index of the shrub density that can be repeated at a variety of heights. It is important to use the same observer, however, as people may differ in this ability. (From Bibby et al., *Bird Census Techniques,* Figure 10.10. Copyright 1992. Reprinted by permission of Academic Press.)

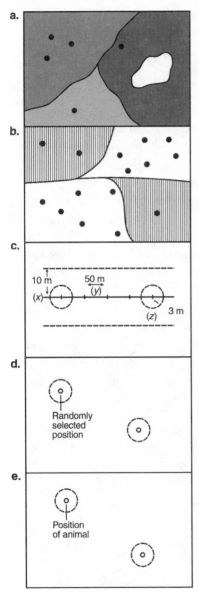

FIGURE 2.6.

Scale of habitat recording for wildlife studies. (a) All vegetation types are mapped without any habitat measurements and the locations of animals are marked on the map (solid dots). This method produces a broad understanding of habitat use, but it is difficult to test relationships statistically. (b) Habitat is subdivided into parcels on the basis of criteria such as vegetation age or plant species composition (white parcels represent recent clear-cutting; shaded parcels indicate old clear-cutting). Animal registrations (solid dots), derived from a mapping census, are allocated to each parcel and compared with quantitatively measured habitat variables. The habitat data from the parcels are produced independently of the mapping and a statistical comparison between the two to test any significant relationships is possible. (c) Habitat variables are recorded in standard sample plots at measured distances along the route of a transect count. This technique produces data on habitat variables in the same position as the transect count and allows the use of multivariate statistical methods to test relationships between animals and habitat variables. x = transect band width, y = measured transect segments, z = sample radius of habitat recording circle. (d) Habitat variables recorded in sample plots around the position of randomly located point counts. This technique produces detailed habitat data in the same position as the point count. Again this technique allows the use of multivariate statistical methods to test relationships between animals and habitat variables. (e) Habitat variables are recorded at the position of a territorial, feeding, or radio-located animal. This technique produces precise habitat data in an area selected by the animal. By recording habitat variables at a random selection of plots within the study area, it is possible to quantify an animal's habitat selection in terms of measured differences in the habitat variables it uses or avoids. (From Bibby et al., *Bird Census Techniques*, Figure 10.1. Copyright 1992. Reprinted by permission of Academic Press.)

these techniques—ranging from mapping of general bird locations to specific assessment of individual habitat use—can be applied in the field.

Lessons

To advance our understanding of habitat relationships will require increased cooperation among wildlife scientists, conservation biologists, and restorationists. Standardization of terminology would be a big step toward such cooperation by promoting use of a common language. The concept of habitat is well established in the scientific and popular literature, for example, yet the term is widely misunderstood and misused. Identifying critical resources that limit the distribution, abundance, survival, and reproductive performance of wildlife is a key factor in designing a restoration project. Moreover, the need to quantify constraints on the acquisition of resources—niche factors—has not been adequately recognized in studies of wildlife ecology and restoration.

Wildlife scientists have expended considerable effort studying efficient means of quantifying animal habitat. Restorationists will be able to accelerate the achievement of their project goals by studying the strengths and weaknesses of previous wildlife/habitat studies. That is: there is no reason to reinvent the wheel or repeat past mistakes. Oversampling or undersampling can be avoided through careful planning and the use of preliminary studies (see Chapter 4).

References

Arthur, W. 1987. *The Niche in Competition and Evolution.* New York: Wiley.

Askins, R. A. 2000. *Restoring North America's Birds.* New Haven: Yale University Press.

Bibby, C. J., N. D. Burgess, and D. A. Hill. 1992. *Bird Census Techniques.* London: Academic Press.

Bissonette, J. A. (ed.). 1997. *Wildlife and Landscape Ecology: Effects of Pattern and Scale.* New York: Springer-Verlag.

Block, W. M., and L. A. Brennan. 1993. The habitat concept in ornithology: Theory and applications. *Current Ornithology* 11:35–91.

Block, W. M., M. L. Morrison, J. Verner, and P. N. Manley. 1994. Assessing wildlife-habitat-relationships models: A case study with California oak woodlands. *Wildlife Society Bulletin* 22:549–561.

Bonham, C. D. 1989. *Measurements for Terrestrial Vegetation*. New York: Wiley.

Bookhout, T. A. (ed). 1994. *Research and Management Techniques for Wildlife and Habitats*. 5th ed. Bethesda, Md.: Wildlife Society.

Bratton, S. P., and A. J. Meier. 1998. Restoring wildflowers and salamanders in southeastern deciduous forests. *Restoration and Management Notes* 16:158–165.

Cook, C. W., and J. Stubbendieck (eds.). 1986. *Range Research: Basic Problems and Techniques*. Denver: Society for Range Management.

Cooperrider, A. Y., R. J. Boyd, and H. R. Stuart (eds.). 1986. *Inventory and Monitoring of Wildlife Habitat*. Denver: USDA Bureau of Land Management Service Center.

Dasmann, R. F., J. P. Milton, and P. H. Freeman. 1973. *Ecological Principles for Economic Development*. London: Wiley.

Daubenmire, R. 1968. *Plant Communities: A Textbook of Plant Synecology*. New York: Harper and Row.

Davis, F. W., and D. M. Stoms. 1996. A spatial analytical hierarchy for GAP analysis. Pages 15–24 in J. M. Scott, T. H. Tear, and F. W. Davis (eds.), *GAP Analysis: A Landscape Approach to Biodiversity Planning*. Bethesda, Md.: American Society for Photogrammetry and Remote Sensing.

Dueser, R. D., and H. H. Shugart Jr. 1978. Microhabitats in forest-floor small mammal fauna. *Ecology* 59:89–98.

Elton, C. 1927. *Animal Ecology*. London: Sidgwick & Jackson.

Forman, R.T.T., and M. Gordon. 1986. *Landscape Ecology*. New York: Wiley.

Green, R. H. 1979. *Sampling Design and Statistical Methods for Environmental Biologists*. New York: Wiley.

Greig-Smith, P. 1983. *Quantitative Plant Ecology*. 3rd ed. Berkeley: University of California Press.

Grinnell, J. 1917. The niche-relationships of the California thrasher. *Auk* 34:427–433.

Grover, M. C. 1998. Influence of cover and moisture on abundances of the terrestrial salamanders *Plethodon cinereus* and *Plethodon glutinosus*. *Journal of Herpetology* 32:489–497.

Hall, L. S., P. R. Krausman, and M. L. Morrison. 1997. The habitat concept and a plea for standard terminology. *Wildlife Society Bulletin* 25:173–182.

Holmes, R. T. 1981. Theoretical aspects of habitat use by birds. Pages 33–37 in D. E. Capen (ed.), *The Use of Multivariate Statistics in Studies of Wildlife Habi-*

tat. Fort Collins, Colo.: USDA Forest Service General Technical Report RM-87.

Hutchinson, G. E. 1957. Concluding remarks. *Cold Spring Harbor Symposium on Quantitative Biology* 22:415–427.

Hutto, R. L. 1985. Habitat selection by nonbreeding, migratory land birds. Pages 455–476 in M. L. Cody (ed.), *Habitat Selection in Birds.* Orlando: Academic Press.

Huxel, G. R., and A. Hastings. 1999. Habitat loss, fragmentation, and restoration. *Restoration Ecology* 7:309–315.

Irwin, L. L., and J. G. Cook. 1985. Determining appropriate variables for a habitat suitability model for pronghorns. *Wildlife Society Bulletin* 13:434–440.

James, F. C. 1971. Ordinations of habitat relationships among breeding birds. *Wilson Bulletin* 83:215–236.

James, F. C., and H. H. Shugart Jr. 1970. A quantitative method of habitat description. *Audubon Field Notes* 24:727–736.

Johnson, D. H. 1980. The comparison of usage and availability measurements for evaluating resource preference. *Ecology* 61:65–71.

King, A. W. 1997. Hierarchy theory: A guide to system structure for wildlife biologists. Pages 185–212 in J. A. Bissonette (ed.), *Wildlife and Landscape Ecology: Effects of Pattern and Scale.* New York: Springer-Verlag.

Laymon, S. A., and R. H. Barrett. 1986. Developing and testing habitat-capability models: Pitfalls and recommendations. Pages 87–91 in J. Verner, M. L. Morrison, and C. J. Ralph (eds.), *Wildlife 2000: Modeling Habitat Relationships of Terrestrial Vertebrates.* Madison: University of Wisconsin Press.

Leopold, A. 1933. *Game Management.* New York: Scribner's.

Litvaitis, J. A., K. Titus, and E. M. Anderson. 1994. Measuring vertebrate use of terrestrial habitats and foods. Pages 254–274 in T. A. Bookhout (ed.), *Research and Management Techniques for Wildlife Habitats.* 5th ed. Bethesda, Md.: Wildlife Society.

MacArthur, R. H. 1968. The theory of the niche. Pages 159–176 in R. C. Lewontin (ed.), *Population Biology and Evolution.* Syracuse: Syracuse University Press.

Morrison, M. L. 1984a. Influence of sample size and sampling design on analysis of avian foraging behavior. *Condor* 86:146–150.

———. 1984b. Influence of sample size and sampling design on discriminant function analysis of habitat use by birds. *Journal of Field Ornithology* 55:330–335.

Morrison, M. L., W. M. Block, and J. Verner. 1991. Wildlife-habitat relationships in California's oak woodlands: Where do we go from here? Pages 105–109 in *Proceedings of the Symposium on California's Oak Woodlands and Hardwood Rangeland.* Washington, D.C.: USDA Forest Service General Technical Report PSW-126.

Morrison, M. L., W. M. Block, L. S. Hall, and H. S. Stone. 1995. Habitat characteristics and monitoring of amphibians and reptiles in the Huachuca Mountains, Arizona. *Southwestern Naturalist* 40:185–192.

Morrison, M. L., B. G. Marcot, and R. W. Mannan. 1992. *Wildlife-Habitat Relationships: Concepts and Applications.* Madison: University of Wisconsin Press.

———. 1998. *Wildlife-Habitat Relationships: Concepts and Applications.* 2nd ed. Madison: University of Wisconsin Press.

Mueller-Dombois, D., and H. Ellenberg. 1974. *Aims and Methods of Vegetation Ecology.* New York: Wiley.

Murray, N. L., and D. F. Stauffer. 1995. Nongame bird use of habitat in central Appalachian riparian forests. *Journal of Wildlife Management* 59:78–88.

Noon, B. R. 1981. Techniques for sampling avian habitats. Pages 42–52 in D. E. Capen (ed.), *The Use of Multivariate Statistics in Studies of Wildlife Habitat.* Fort Collins, Colo.: USDA Forest Service General Technical Report RM-87.

Odum, E. 1971. *Fundamentals of Ecology.* 3rd ed. Philadelphia: Saunders.

Palmer, M. A., R. F. Ambrose, and N. L. Poff. 1997. Ecological theory and community restoration ecology. *Restoration Ecology* 5:291–300.

Peters, R. H. 1991. *A Critique for Ecology.* Cambridge: Cambridge University Press.

Pulliam, H. R. 1988. Sources, sinks, and population regulation. *American Naturalist* 132:652–661.

Reinert, H. K. 1984. Habitat separation between sympatric snake populations. *Ecology* 65:478–486.

Roth, R. R. 1976. Spatial heterogeneity and bird species diversity. *Ecology* 57:773–782.

Schooley, R. L. 1994. Annual variation in habitat selection: Patterns concealed by pooled data. *Journal of Wildlife Management* 58:367–374.

Schreuder, H. T., T. G. Gregoire, and G. B. Wood. 1993. *Sampling Methods for Multiresource Forest Inventory.* New York: Wiley.

Scott, J. M., F. Davis, B. Csuti, R. Noss, B. Butterfield, C. Grives, H. Anderson, S. Caicco, F. D'Erchia, T. Edwards Jr., J. Ulliman, and R. G. Wright. 1993. Gap analysis: A geographical approach to protection of biodiversity. *Wildlife Monographs* 123:1–41.

U.S. Fish and Wildlife Service (USFWS). 1981. *Standards for the Development of Habitat Suitability Index Models.* Ecological Services Manual 103. Washington, D.C.: Government Printing Office.

VanderWerf, E. A. 1993. Scales of habitat selection by foraging 'elepaio in undisturbed and human-altered forests in Hawaii. *Condor* 95:980–989.

Van Horne, B. 1983. Density as a misleading indicator of habitat quality. *Journal of Wildlife Management* 47:893–901.

Verner, J., M. L. Morrison, and C. J. Ralph (eds.). 1986. *Wildlife 2000: Modeling Habitat Relationships of Terrestrial Vertebrates.* Madison: University of Wisconsin Press.

Wecker, S. C. 1964. Habitat selection. *Scientific American* 211:109–116.

Welsh, H. H., Jr., and A. J. Lind. 1995. Habitat correlates of Del Norte salamander, *Plethodon elongates* (Caudata: Plethodontidae), in northwestern California. *Journal of Herpetology* 29:198–210.

Wenny, D. G., R. L. Clawson, J. Faaborg, and S. L. Sheriff. 1993. Population density, habitat selection, and minimum area requirements of three forest-interior warblers in central Missouri. *Condor* 95:968–979.

Wiens, J. A. 1972. Anuran habitat selection: Early experience and substrate selection in *Rana cascadae* tadpoles. *Animal Behavior* 20:218–220.

———. 1984. The place of long-term studies in ornithology. *Auk* 101:202–203.

———. 1989a. *The Ecology of Bird Communities.* Vol. 1: *Foundations and Patterns.* Cambridge: Cambridge University Press.

———. 1989b. *The Ecology of Bird Communities.* Vol. 2: *Processes and Variations.* Cambridge: Cambridge University Press.

Willson, M. F. 1974. Avian community organization and habitat structure. *Ecology* 55:1017–1029.

Wootton, J. T., and D. A. Bell. 1992. A metapopulation model of the peregrine falcon in California: Viability and management strategies. *Ecological Applications* 2:307–321.

Chapter 3

Historic Assessments

One of the first steps in designing a restoration project is to establish a time period for replicating all or part of an ecosystem (Egan and Howell 2001). This planning process should include evaluation of the historic animal communities. (See Swetnam et al. 1999 for a review.) As we shall see throughout this book, simply providing a general vegetation type or plant association is unlikely to yield the necessary habitat components to allow occupancy by many animal species.

Restoring an area to match some preexisting condition is difficult unless you have data on historic conditions (Figure 3.1). The goal of this chapter is to present techniques useful in reconstructing the historic assemblage of animals in an area by describing methods for gathering historic data on animal occurrences and determining the uncertainty associated with historic data. Case studies on the use of such techniques are discussed as well. For a detailed discussion see Morrison (2001).

Background

Most animal species currently occupying the earth are survivors of the abiotic and biotic influences of the Pleistocene. The Pleistocene epoch, which began 2 or 3 million years ago, is thought to have ended about 10,000 years before the present. The Pleistocene was characterized by a series of advances

FIGURE 3.1.
A pine stand, thinned to replicate natural fire regimes, is an example of restorationists attempting to replicate "natural" conditions. (Photo courtesy of Bruce G. Marcot.)

and retreats of continental ice sheets and glaciers. Our Recent epoch is, in fact, probably another interglacial period of the Pleistocene (Cox and Moore 1993). Thus the current distribution and abundance of animal species existing today can be linked to the geological events of the Pleistocene. The retreat of the ice sheet allowed occupation of vast areas either by species already adapted to the newly developing vegetation or by those able to adjust to the new environmental conditions. The present range of many taxa was probably reached in the early Holocene, but some survived in refugia until the late Holocene (Elias 1992). The influence of glacial-interglacial cycles on species' geographic ranges has been substantial. (See, for example, the review by Gutierrez 1997.)

In the American Southwest, for example, both post-Pleistocene dispersal and subsequent colonization as well as vicariant events (the distribution that results from the replacement of one member of a species pair by another) and subsequent extinction have influenced the current assemblages of mammals. Davis et al. (1988) stress that the degree to which each process influenced animal distribution should be considered in explaining current faunal composition. When Johnson (1994) studied the range expansion of 24 species of birds in the contiguous western United States, he found that climatic information from the region indicated wetter and warmer summers in recent decades, which he related to the range expansions. The restorationist can use such phenomena to identify potential animal communities in the project area.

Predicting the exact species composition of a locality is extremely difficult. We can understand this by reviewing the definition of habitat (see Chapter 2). Habitat is a species-specific concept that includes more than vegetation. Identifying the vegetation type of a locality is not the same as identifying the many species-specific habitats of the locality. Without observations or specimens, we can never be positive that the species of interest ever occurred on the site. We can, however, assemble lists of probable species occurrences based on such evidence as distance to the nearest verified record of the species. The more thoroughly we can describe the site's historic vegetative and environmental conditions (presence of permanent water, for instance, and soil conditions), the more complete our list of potential species will be.

Kessel and Gibson (1994) conclude that changes in a region's animal communities may be categorized as those we believe are real changes, those that merely reflect our increased knowledge of the animal community, those that may be natural long-term fluctuations, and those attributable to confused species identifications. Today it is almost impossible to tell whether the changes we perceive are real directional changes or just fluctuations—and some species fall into more than one of these categories. For most species we have only sporadic, vague comments on status in the historical literature or data too recent or too incomplete to offer a basis for reconstructing past communities and changes through time.

Techniques

There are many sources of data that can be consulted to reconstruct a list of species formerly using an area. In this section I outline several major techniques, including existing data, specimens housed in museums, the fossil record, and the abundant natural history literature. Together these sources often provide a rich background on the ecology of an area of interest.

Data Sources

The USGS–Biological Resources Division (formerly managed within the U.S. Fish and Wildlife Service) coordinates the nationwide Breeding Bird

Survey (BBS). Initiated in 1965, the BBS consists of more than 2000 randomly located permanent routes established along secondary roads throughout the continental United States and southern Canada that are surveyed annually during the height of the breeding season, usually in June. Each route is 25 miles (40 km) long and consists of 50 stops spaced at 0.8-km intervals (Robbins et al. 1986).

The National Audubon Society (NAS) coordinates an annual bird count in December. Known as the Christmas Bird Count (CBC), this one-day effort is conducted by volunteers in a 15-mile radius (24 km) of a chosen location (usually a city, wildlife refuge, or any other area of interest). Begun in 1900, the count has grown into a valuable database for long-term monitoring of population trends. CBC data have been summarized by the USGS-BRD and by various independent researchers (such as Wing 1947). Similar count data were recorded in the *Canadian Field Naturalist* from 1924 through 1939. Counting areas were initially concentrated in the eastern United States. Since the 1950s, however, counting areas have become more common throughout the continental United States. The increasing coverage and density of counting areas are expanding our knowledge of bird distribution and will lead to more accurate numbers in the future. The data are published annually by NAS in *National Audubon Field-Notes* (formerly *American Birds*). The BBS and CBC data represent a relatively recent list of the distribution and abundance of birds. Although these data do not provide historical information for many locations at this time, they might be available for your project area.

There are many published papers and reports that you can use to reconstruct the fauna of a region. These are the result of natural history surveys conducted during the late 1800s and early 1900s. In every region of North America there are journals that emphasize natural history reports. A thorough literature survey will allow you to reconstruct the fauna occurring in the location prior to the advent of intensive human development. Remember, of course, that natural catastrophes (floods, fires, tornadoes, drought) also influence the local distribution of species substantially. As we shall see, extending the results of localized surveys to other sites reduces the confidence you can place in the reconstruction.

Museum Records

Natural history collections are housed in a variety of private and public universities, museums, and research organizations. Usually these collections were assembled to characterize the fauna of a region. Each specimen is accompanied by an original paper tag that lists the date and location of collection, the collector, the species, and perhaps a few natural history notes.

Although museums tend to concentrate on species within their geographic region, many of the larger ones have gathered specimens from throughout North America and even the world.

Bird eggs are sometimes housed in ornithological collections. The science of studying eggs (oology) was extremely popular in the early 1900s, and several oological journals were published. Egg collections have been valuable in the analysis of natural history parameters (clutch size, breeding phenology), breeding distributions, and eggshell thinning. The data slip accompanying the egg sets usually includes clutch size, nest location (height, plant substrate), and related data. Kiff and Hough (1985) present a detailed summary of specimen holdings, geographic coverage, and related information on egg collections in North America. The premier collection is held by the Western Foundation of Vertebrate Zoology (WFVZ), Camarillo, California. The WFVZ can be contacted for additional information on accessing oological collections.

Warning: An increasing number of museums are entering their original data into computer databases. This is a favorable trend as it helps the museum to manage the collection and answer questions regarding specimen holdings. But users must be aware that errors may occur during the transcription of original data. Moreover, few programs transcribe all the information—including natural history notes—into the database. Thus the user would be wise to request a computer printout of specimen holdings (sorted by date and location) of the species of interest beforehand and then ask for photocopies of the original data slips. Moreover, the identification of specimens in many collections has never been verified. Thus it is often necessary to visit the collection personally to confirm identification, request a loan of critical specimens, or ask for independent verification of the specimen by a local expert. The presence of a specimen in a collection simply indicates that the species was present at the time of collection. You cannot use the absence of a species in a collection to conclude that the species did not occur at the time the collecting was under way. Yet people have misused museum collections by interpreting the lack of specimens of a species of interest (say, an endangered species) as indicating lack of occurrence. The novice should consult people experienced with such matters before using the information.

Those contacting museums for specimen information should be aware of the serious underfunding of most institutions. Thus even academic users should offer to pay for photocopies of data or for the labor involved in accessing and printing computer databases. Those who use museum data should clearly discuss any limitations and biases in the records. Requests for specimens for personal use (to verify identification, for example, or take

measurements) should be kept to a minimum. No only do requests for specimens place a serious workload on museum staff, but handling and shipping
specimens alters their measurements and shortens their useful lifetime.

Fossils and Subfossils

The fossil record can sometimes be used to reconstruct the former range of
species. Harris (1993) reconstructed the succession of microtene rodents
from the mid- to late-Wisconsin period of the Pleistocene in New Mexico,
for example, and Goodwin (1995) reconstructed the Pleistocene distribution
of prairie dogs (*Cynomys* spp.). Reconstructing the biogeographic history of
a species reveals the changing distribution of its habitat. Moreover, such a
reconstruction indicates the historical biogeography of other recent species
that use the same general conditions. Hafner (1993), for example, used the
Nearctic pikas (*Ochotona princeps* and *O. collaris*) as biogeographic indicators

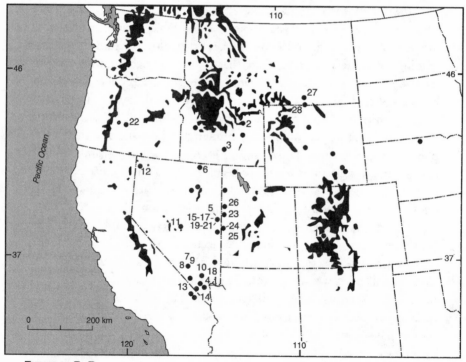

FIGURE 3.2.
Distribution (shaded areas) of extant pikas and late Pleistocene-Holocene fossil records
(solid dots) in western North America. (From D. J. Hafner, "North American Pika as a
Late Quarternary Biogeographic Indicator Species," Figure 1. *Quarternary Research*
39:373–380. Copyright 1993, Academic Press.)

of cool, mesic, rocky areas. Fossil pikas have been found far from extant populations, especially in Nevada (Figure 3.2). Such reconstructions not only help us to understand why species change in distribution but also lend insight into the critical factors limiting current distribution.

Subfossil remains have been used to reconstruct the environment as well. *Subfossils* are unmineralized remains that may be only a few hundred years old. Subfossils are often found in caves, mines, woodrat (*Neotoma* spp.; "packrat") middens, and in numerous rocky crevasses. Ashmole and Ashmole (1997), for example, used subfossils to help them reconstruct the prehistoric ecosystem of Ascension Island in the equatorial Atlantic. (See also Olson 1977.)

To identify fossil or subfossil remains you will need reference specimens for comparison with the unknown items. Because we are restricting our analyses to the Holocene and forward in time, natural history museums are likely to contain adequate specimens for comparison. Knowledge of vertebrate morphology is necessary to speed the identification process, but those with undergraduate training in wildlife science or zoology can usually perform the analyses.

Kay (1998) reviewed archaeological and other data to reconstruct the pre-Columbian ungulate and predator (including human) composition of North America. He concludes that Native Americans were the ultimate predator structuring North American ecosystems from about 12,000 years ago to about 1870, especially in the western United States and Canada. Such work has serious implications for restoration by establishing goals for the composition of animal assemblages—and thus vegetation types and successional patterns. Kay has showed, for example, that of nearly 60,000 ungulate bones unearthed at more than 400 archaeological sites in the United States and Canadian Rockies, fewer than 3 percent were elk and only about 10 percent were bison. Using this along with other evidence, he concludes that ungulates were kept at low numbers (relative to the present) by Native American hunters, which, in combination with human-started fires, had serious implications for the shape of the vegetative landscape.

Literature

Assessing changes in animal communities is a topic of particular interest to conservation biologists trying to separate human-induced changes from those caused by other environmental factors. These are usually termed "natural" changes, although this is a misnomer given that humans evolved on this planet. Thus there is much literature on determining past animal communities. Documenting changes requires a temporal baseline against which subsequent records can be compared and evaluated. Prior to the early 1900s such baselines were lacking for most regions. From the early 1900s through the 1950s, however, regional surveys accompanied by mass collecting established reasonably precise ranges for many species.

Power (1994) has published a good example of assembling historic avifaunal records. Using historic papers beginning in the 1850s (from the U.S. Pacific Railroad Survey, for example) and continuing through the early 1900s, he reconstructed the distribution and abundance of birds of the coastal islands of California. Other useful papers have appeared in *Proceedings of the California Academy of Sciences, Pasadena Academy of Science Publication, Proceedings of the National Academy of Sciences, Pacific Coast Avifauna*, and the journals *Condor* and *Auk*. Both national and regional publications are a source of substantial information.

The National Audubon Society has published a compilation of bird observations submitted by the public throughout the 1900s. These observations are divided by geographic region of North America and summarize seasonal (winter, summer, fall) observations. Experts on the birds of each region then compile and verify the records submitted by field observers. The narrative accompanying many of the observations explains the status of the species in the region. Although these data are useful for interpreting the status of rarer species of a region, there are few long-term data on site-specific changes in occurrence.

Field notes, journals, and other written records are contained in various publications (books, journal articles) or are housed in their original form in museums. These written records often contain species-specific information on distribution, abundance, breeding status, and other natural history notes. They can help you to reconstruct the animals present in the region along with the general environmental conditions.

Uncertainty

Without direct observations of a species' presence and activity, we cannot assign a specific probability to our reconstructions of historic animal assemblage. Even if specimens exist for the locality of interest, we seldom know the species' status on the site. Even the presence of an adult of a nonmigratory species at the location does not mean it was actually resident there—though naturally we would assign a higher probability of its residence on the site compared to a specimen for a migratory species. Thus we can assess the quality of our historic reconstruction by assigning probabilities of certainty to each data source. Such assignment of certainty is relative and qualitative. Factors to consider include:

- Age of the data source: Old data sources are not necessarily unreliable. The validity of a source, however, becomes increasingly difficult to determine with age.
- Distance from the data source: The occurrence of a species in the vicinity of a restoration site might be used to infer that the species once existed on the site. And the further a historic record of occurrence is from the restoration site, the less likely it is that the species occurred on the site. In this evaluation, information on dispersal distances of the species of interest is essential.
- Quantity and quality of data sources: Here you have to consider one record versus numerous records, records from a brief time period versus samples across time, actual specimens versus visual observations, completeness of the data record, and reputation of the data source.

The uncertainty of each conclusion needs to be discussed, and all assumptions must be stated.

Case Studies

To develop a wildlife habitat restoration plan for the Sweetwater Regional Park, San Diego County, California, Morrison et al. (1994a, 1994b) analyzed the past and current distribution and abundance of vegetation and wildlife. They compared new survey results for amphibians, reptiles, mammals, and birds with historical data obtained from specimens housed at the San Diego Natural History Museum. Because intensive residential and commercial development began in the mid-1970s, "historical" was defined as pre-1975 collections. Literature sources were used to supplement the museum records. In summary, their work indicated a substantial loss of native amphibians and reptiles, including 4 amphibians, 3 lizards, and 11 snake species. The small-mammal community was depauperate and dominated by the nonnative house mouse (*Mus musculus*) and the native western harvest mouse (*Reithrodontomys megalotis*). There was an apparent net loss of 13 mammal species: 9 insectivores (such as shrews, *Sorex*) and rodents, 1 rabbit, and 3 large mammals. There was an absolute loss of 18 bird species and a gain of 6 species. They used these data to help them develop a plan for restoring the plant and animal communities.

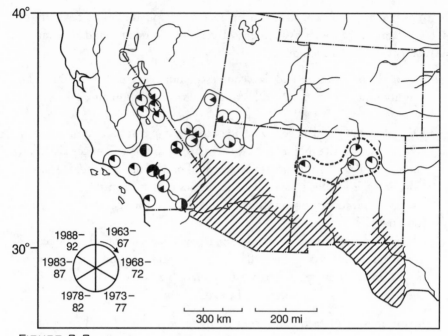

FIGURE 3.3.
The pattern of diagonal lines shows the approximate breeding range of the western form of the summer tanager (*Piranga rubra cooperi*) as of the mid-1950s to early 1960s in the American Southwest. Localities of range expansion by pioneers and colonists in the last three decades are denoted by the symbols and dashed line. (From N. K. Johnson, "Pioneering and Natural Expansion of Breeding Distributions in North America," Figure 7. *Studies in Avian Biology* 15:27–44. Copyright 1994.)

Johnson (1994) calculated distribution changes by comparing a midcentury baseline with subsequently published regional avifaunal compilations, including newer editions of the *Check-list*. For detailed information on pioneers and extralimital nesting over the last three decades, he tallied nesting season records cited in *Audubon Field-Notes* and *American Birds*. Johnson emphasized extralimital late spring records because they often point to pioneering and imminent summer residence. He defined pioneering as the presence of a singing male or a pair in appropriate breeding habitat. As an example, Johnson reconstructed the steadily expanding range of the summer tanager (*Piranga rubra cooperi*) in the American Southwest (Figure 3.3). His analysis identifies the species' range as of the midcentury and traces its steady northward and westward expansion. For this species and most of the others he analyzed, Johnson concluded that recent range expansions were due to climatic changes (increased summer moisture and higher mean summer temperatures)—not responses to human activity. His analysis allows restoration planners to understand the history of a species in the region of interest and establish priorities regarding development of vegetation communities and specific plant associations.

Lessons

Reconstructing the likely animal species assemblages of a proposed restoration site is a valuable way of developing project goals. But the historic presence of a species does not mean it could reoccupy the site even if all the habitat conditions were restored. Many species have certain minimum area requirements that cannot be met on small restoration sites. Further, the conditions on land surrounding the restoration site may be unsuitable to allow recolonization. Thus you should develop your restoration goals for wildlife in light of both historic possibilities and current realities.

Gathering historical records is a time-consuming task. Often animal specimens are scattered among many museums, sometimes in distant locations. Additionally, many records of animal occurrences are contained in old (often pre-1900) journals and unpublished field notes. Although specimens are usually reliable, the validity of such records is difficult to determine.

The literature is full of species accounts, however, including many natural history anecdotes that can be helpful in piecing together the puzzle of what once occurred in an area. A major lesson here is to conduct a thorough data search in the early stages of a restoration project so that you can determine the history of the animals and vegetation that formerly occupied the area.

References

Ashmole, N. P., and M. J. Ashmole. 1997. The land fauna of Ascension Island: New data from caves and lava flows, and a reconstruction of the prehistoric ecosystem. *Journal of Biogeography* 24:549–589.

Cox, C. B., and P. D. Moore. 1993. *Biogeography: An Ecological and Evolutionary Approach.* 5th ed. Boston: Blackwell.

Daubenmire, R. 1968. *Plant Communities: A Textbook of Plant Synecology.* New York: Harper & Row.

Davis, R., C. Dunford, and M. V. Lomolino. 1988. Montane mammals of the American Southwest: The possible influence of post-Pleistocene colonization. *Journal of Biogeography* 15:841–848.

Egan, D., and E. A. Howell (eds.). 2001. *The Historical Ecology Handbook: A Restorationist's Guide to Reference Ecosystems.* Washington, D.C.: Island Press.

Elias, S. A. 1992. Late Quaternary zoogeography of the Chihuahuan Desert insect fauna, based on fossil records from packrat middens. *Journal of Biogeography* 19:185–197.

Goodwin, H. T. 1995. Pliocene-Pleistocene biogeographic history of prairie dogs, *Cynomys* (Sciuridae). *Journal of Mammology* 76:100–122.

Gutierrez, D. 1997. Importance of historical factors on species richness and composition of butterfly assemblages (Lepidoptera: Rhopalocera) in a northern Iberian mountain range. *Journal of Biogeography* 24:77–88.

Hafner, D. J. 1993. North American pika (*Ochotona princeps*) as a late Quaternary biogeographic indicator species. *Quaternary Research* 39:373–380.

Harris, A. H. 1993. Wisconsin and pre-pleniglacial biotic changes in southeastern New Mexico. *Quaternary Research* 40:127–133.

Johnson, N. K. 1994. Pioneering and natural expansion of breeding distributions in western North America. *Studies in Avian Biology* 15:27–44.

Kay, C. E. 1998. Are ecosystems structured from the top-down or bottom-up?: A new look at an old debate. *Wildlife Society Bulletin* 26:484–498.

Kessel, B., and D. D. Gibson. 1994. A century of avifaunal change in Alaska. *Studies in Avian Biology* 15:4–13.

Kiff, L. F., and D. J. Hough. 1985. *Inventory of Bird Egg Collections of North America.* Norman: American Ornithologists' Union and Oklahoma Biological Survey.

King, A. W. 1998. Hierarchy theory: A guide to system structure for wildlife biologists. Pages 185–212 in J. A. Bissonette (ed.), *Wildlife and Landscape Ecology: Effects of Pattern and Scale.* New York: Springer-Verlag.

Morrison, M. L. 2001. Techniques for discovering historic animal assemblages. Pages 295–315 in D. Egan and E. A. Howell (eds.), *The Historical Ecology Handbook: A Restorationist's Guide to Reference Ecosystems.* Washington, D.C.: Island Press.

Morrison, M. L., T. A. Scott, and T. Tennant. 1994a. Wildlife-habitat restoration in an urban park in southern California. *Restoration Ecology* 2:17–30.

———. 1994b. Laying the foundation for a comprehensive program of restoration for wildlife habitat in a riparian floodplain. *Environmental Management* 18:939–955.

Olson, S. L. 1977. Additional notes on subfossil bird remains from Ascension Island. *Ibis* 119:37–43.

Power, D. M. 1994. Avifaunal change on California's coastal islands. *Studies in Avian Biology* 15:75–90.

Robbins, C. S., D. Bystrak, and P. H. Geissler. 1986. *The Breeding Bird Survey: Its First Fifteen Years, 1965–1979.* Research Publication 157. Washington, D.C.: U.S. Fish and Wildlife Service.

Swetnam, T. W., C. D. Allen, and J. L. Betancourt. 1999. Applied historical ecology: Using the past to manage for the future. *Ecological Applications* 9:1189–1206.

Wing, L. 1947. Christmas census summary, 1900–1939. Pullman: State College of Washington. Mimeograph.

Chapter 4

A Primer on Study Design

We are increasingly faced with concerns over destructive human use of the world: pollutants, extinctions, the introduction of exotic plants and animals, fragmentation of habitats, and a host of other impacts. Thus we are faced with predicting the response of flora and fauna to environmental change within a context of increasingly complex interactions due to human activity. We are not very good at predicting the response of animals to "natural" environmental changes, however, let alone making these predictions with the added complexity of planned or unplanned human intervention.

Clearly, then, we must adhere to scientific methods that provide a known and repeatable level of knowledge if we are to advance our understanding of the environment and thus be in a better position to predict the consequences of our actions. Clearly we must avoid basing our management decisions on preconceived notions and built-in biases that derive from poorly conceived studies or indeed no studies at all. All science requires adherence to rigorous scientific methods; only good science should ever be applied. Yet ecology as a discipline has failed to provide reliable knowledge, not because it is inherently flawed, but because its practitioners have failed to treat it as rigorous science (Romesburg 1981; Peters 1991).

My goal for this chapter is to summarize the basic principles of study design—that is, how to gain reliable information for a restoration study. Although my focus is on wildlife habitat, this chapter could easily be adapted to the study of any plant or animal. For a detailed discussion see Morrison (1997) and Morrison et al. (1998:chap. 4).

Scientific Methods

Knowledge is discovered through the application of scientific methods. There is no single scientific method. Rather, different methods are suitable for differing purposes. Romesburg (1981) lists three main scientific methods: induction, retroduction, and hypothetico-deduction. *Induction* applies to the finding of associations between classes of facts—for example, making the observation that certain animals are usually seen within the interior of an oak (*Quercus* spp.) woodland and then correlating the abundance of these animals with distance from the woodland edge. The more times (trials) this association is made, the more reliable it becomes. Thus induction is reliable if done properly, but it does not provide insight into the *process* causing the relationship. Management decisions based on induction often fail because they try to extend the relationship well beyond its original intent—for example, by applying the relationship developed in the oak woodland to a coniferous forest.

Retroduction refers to the development of research hypotheses about processes that are explanations for observed relationships. In the oak woodland example, we might hypothesize that the association was due to the inability of predators to penetrate deeply into the woodland, thus providing protection to animals in the interior portions. Such explanations often yield unreliable information, however, because there may well be many possible explanations. Although our initial explanation seems reasonable, the relationship might actually be due to favorable microclimatic conditions in the woodland interior. Thus a management decision to begin predator control in order to enhance the species of interest would fail. (The correct decision would probably be to increase the woodland's size.)

This problem with retroduction (and induction as well) brings us to Romesburg's third category: the *hypothetico-deductive* (H-D) method. The H-D method complements retroduction in that it starts with a hypothesis—usually developed through retroduction—and then makes testable predictions about other classes of facts that should be true if the research hypothesis is true. Thus H-D is a method of determining the reliability of our ideas. In the woodland example, we would design an experiment to test the influence of distance from edge, microclimatic conditions, predator abundance, and perhaps several measures of woodland structure on some measure of the animal population (abundance, for instance, or breeding success).

Successful restoration, then, involves understanding the basis for an ecological relationship. Careful evaluation of the source of the information (journal article, book chapter, word of mouth) and how the information was obtained will indicate the reliance you should place on it. Clearly a rigorous study based on the H-D method that identifies critical resource items will give you more confidence than an inductive idea. Most studies in ecology, however, are based on the inductive or retroductive methods. This is one reason why researchers have been calling for studies that seek to determine *why* relationships are occurring (Gavin 1989). If the underlying processes are understood, we have a better chance of applying results obtained in one area (or time) to another. This does not diminish the usefulness of inductive-retroductive studies. It simply means that we should not expect them to do more than they are designed to do. And we should avoid the temptation to base our conclusions and restoration plans on information that may be biased and unreliable.

Monitoring as Research

A common misconception among resource managers is that monitoring is different from research and therefore requires less rigor in its design and implementation. This misconception has led to many poorly designed studies and bad management decisions (Morrison and Marcot 1994). As defined by Green (1979:68), a monitoring study is meant to detect a change from the present conditions. The resulting data then provide a baseline against which future changes (impacts) can be measured. But the precision at which such an impact can be measured depends on the rigor of the monitoring study. Just "going out there and collecting some data" in the vague hope that the information might turn out to be useful is foolish.

In developing a monitoring study, then, researchers must decide beforehand what level of change will be considered an impact. (See Figure 4.1.) This decision will determine the sampling intensity (number of study plots, for example) and frequency (say, time between sampling). Suppose you are interested in tracking the change in the abundance of an animal that is the focus of a restoration project. First you must determine the amount of change considered to be biologically meaningful. Then you can take steps to develop the sampling design and protocol needed to detect this change with

FIGURE 4.1.
Advances in technology allow monitoring of animal movements, habitat use, and mortality in wide-ranging species. Here researchers are attaching a satellite radio collar to a female bighorn sheep in southern Arizona. (Photo courtesy of Paul R. Krausman.)

the desired degree of certainty. The design and implementation of monitoring studies are covered in Chapter 6. For a detailed presentation of monitoring see also Morrison et al. (2001).

Principles of Study Design

In what is certainly a classic work on study design, Green (1979) developed basic principles of sampling design and statistical analysis of relevance to environmental studies. These principles, discussed by Morrison (1997) in reference to plant removal and restoration, are summarized in Table 4.1. Here I want to discuss several of Green's principles with regard to wildlife/habitat restoration.

A study's goals flow from the initial statement of the problem (will native rodents, for example, eat the seeds of an exotic plant?) to the statement of the null hypothesis (the density of exotic seeds, for example, does not significantly reduce the mass of a certain native rodent species). It is absolutely critical that this step be well developed, or everything that follows might be for naught. In determining the goals for your study, it is essential to estab-

TABLE 4.1. Ten Principles of Study Design

1. Be able to state concisely to someone else the question you are asking. Your results will be as coherent and as comprehensible as your initial conception of the problem.
2. Take replicate samples within each combination of time, location, and any other controlled variable. Replication is essential.
3. Take an equal number of randomly allocated replicate samples for each combination of controlled variables.
4. To test whether a condition has an effect, collect samples where the condition is present and where the condition is absent but everything else is similar. An effect can be shown only by comparison with a control.
5. Conduct preliminary sampling to provide a basis for evaluation of sampling designs and options for analysis.
6. Verify that your are in fact sampling the population you think you are sampling—and with equal and adequate efficiency over the range of sampling conditions likely to be encountered.
7. If the area to be sampled has a large-scale environmental pattern, break the area up into homogeneous subareas and allocate samples to each in proportion to the size of each subarea.
8. Verify that your sample unit is appropriate to the sizes, densities, and spatial distributions of the organisms you are sampling. Then estimate the number of replicate samples you will need to obtain the desired precision.
9. Test your data to determine whether the error variation is homogeneous, normally distributed, and independent of the mean.
10. Having chosen the best statistical methods to test your hypothesis, stick with the result. An unexpected or undesired result is not a valid reason for rejecting the method and hunting for a "better" one.

Source: R. H. Green, *Sampling Designs and Statistical Methods for Environmental Biologists.* Copyright 1979. Reprinted by permission of John Wiley & Sons, Inc.

lish the spatial and temporal applicability of the results. If your goal is to determine the best means of reducing an exotic that has a wide distribution, for example, then study plots restricted in space and time are unlikely to have broad applicability because of variations in environmental conditions. Such considerations will determine the spatial and temporal nature of your sampling.

The confidence you desire in the results must be determined during the planning stage also. Asking whether a reduction in canopy cover reduces breeding success by 50 ± 10 percent requires much more rigorous sampling than asking whether it reduces it by 50 ± 25 percent. Similarly, asking whether canopy cover *changes* breeding success is different from asking whether it *reduces* success. There is an obvious interplay between the question you ask, the sample size you need to answer it reliably, and thus the generality you can expect from your results.

Conducting a pilot study is an important but often overlooked step. Pilot studies are needed when you have little notion of the best sampling design, sampling intensity, or even sampling methods. Green (1979:31) warns: "Those who skip this step because they do not have enough time usually end up losing time." Preliminary work allows you to verify that your sampling procedures are actually sampling the segment of the population or environment that you intended. It also allows you to evaluate sample sizes and modify your sampling techniques before you get locked into a particular methodology. This step is usually skipped, however, and certainly leads to more wasted time and money than any other stage in a research study. Before charging into grid establishment, you should first "study the study." Few studies are so unique that the methods have never been used before. Thus most of the data collected during the pilot study will be available for analysis with the overall data set.

If there are variations in sampling from area to area, comparison will be biased. In studies that require capturing animals, it is well established that different species—as well as different ages and sexes within species—have differential capturability. Often you will need different sampling tools and methods to sample the species composition and vegetative profile of an area adequately. If a sampling method is applied consistently across all study locations, bias may not be a problem. But even the appearance of bias in a study can greatly diminish the importance that others will attach to your results. You should identify potential biases in the planning stages and then take steps to remove their influence. As Green (1979:38–39) points out, the spatial distribution of organisms is important in establishing a sampling unit. Most ecologists simply gather as much data as they can—somehow hoping that their efforts will be adequate to persuade peer reviewers to publish their work. Given limited budgets, however, neither oversampling nor undersampling is a productive use of your time.

Green (1979:44) contends that environmental studies fall into two general categories: either they ignore the fact that there are methodological and statistical assumptions at all, or they are paranoid about them and rely on nonparametric procedures. Green argues that the method's assumptions should be understood at the time it is chosen, that the likelihood and consequences of violation should be addressed, and that use of the method should then proceed with awareness of the risks and the possible remedies. Assumptions such as normality and homogeneity of variances are seldom met (especially for multivariate data). Violation of assumptions does not always negate the use of parametric tests, however, because most are robust to the violation of assumptions—given an adequate sample size. There is no reason to think that biological phenomena will be linear or have equal distributions among groups; such conditions are abnormal. Green (1979:43–63), along with most general statistic texts, discusses transformations and related topics in detail.

There are several prerequisites for an optimal study design. First: the impact (a general term used for change—for example, restoration-related construction or treatment) must not have occurred, so that before-impact baseline data can provide a temporal control for comparing with post-impact data. Second: the type of impact and the time and place of occurrence must be known. And third: controls must be available (Figure 4.2, sequence 1). As noted by Green, therefore, the optimal design is an area-by-time factorial design in which evidence for an impact is a significant area-by-time interaction. Once the prerequisites for an optimal design are met, the choice of sampling design and statistical analysis should be based on your ability to test the null hypothesis that any change in the impacted area does not differ from the control; your ability to relate to the impact any demonstrated change unique to the impacted area; and to separate effects caused by naturally occurring variation unrelated to the impact (Green 1979:71). Thus an optimal restoration project gathers baseline data prior to implementation of

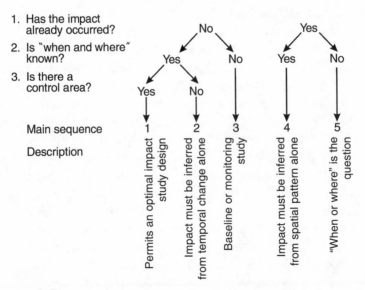

FIGURE 4.2.
The decision key to the "main sequence" categories of environmental studies. (From R. H. Green, *Sampling Designs and Statistical Methods for Environmental Biologists*, Figure 3.4. Page 72. Copyright 1979. Reprinted by permission of John Wiley & Sons, Inc.)

the plan and has adequate control areas to ensure that you learn the true response of animals (and plants) to your restoration activities.

Optimal and Suboptimal Study Designs

Sometimes it is impossible to meet the criteria for an optimal design. Perhaps impacts occur unexpectedly, there is not enough time or funding to implement the design, or reference areas are unfeasible or logistically impossible. In such cases, you may have to use a suboptimal design. If no control areas are possible (Figure 4.2, sequence 2), then you must infer the significance of the change from temporal changes alone. If the location and timing of the impact are not known (fire, flood, disease), then the study becomes a baseline or monitoring study (Figure 4.2, sequence 3). If your study is properly planned spatially, then nonimpacted areas will probably be available to serve as controls if and when the impact occurs. Again we see why monitoring studies are certainly research and permit the development of a rigorous experimental analysis if properly planned.

Often, however, the impact occurs without any planning by the land manager. This common situation (Figure 4.2, sequence 4) means that you must infer the impact's effects from areas differing in the degree of impact. For restoration, this situation often occurs because the project goal is to repair some environmental damage. Sometimes an impact is known to have occurred (Figure 4.2, sequence 5), but the time and location are uncertain (the discovery of a toxin in a plant, for example).

As noted earlier, Green (1979) developed many of the basic principles of environmental sampling design. Most notably, his Before-After-Control-Impact (BACI) design (sequence 1) is the standard upon which most current designs are based. It has been noted, however, that differences may occur between two times of sampling without being caused by the impact of interest (Hurlbert 1984; Underwood 1994). Several researchers (Bernstein and Zalinski 1983; Stewart-Oaten et al. 1986) have expanded BACI to include replicated times of sampling both before and after the impact. Moreover, Osenberg et al. (1994) have developed a BACI design with paired sampling (BACIP). BACIP requires paired (simultaneous or nearly so) sampling several times before and after the impact at both the control and impacted site. The measure of interest is the difference ("delta" in statistical terms) in a parameter value between the control and impacted site as assessed on each sampling date. (See Morrison et al. 2001 for details.)

Repeatedly sampling from the same location, however, raises concerns regarding pseudoreplication—a sampling problem popularized by Hurlbert (1984). Ecologists commonly sample from the same location at different times (weekly, monthly) and consider each sampling event an independent sample. But such a "design" usually inflates the degrees of freedom artificially, leading to erroneous statistical analysis. Even if a case can be made for repeatedly sampling from the same location, this temporal modification of BACI does not solve the problem of lack of spatial replication. As every ecologist knows, organisms often have different temporal trends in different locations (Underwood 1994). Although replicated impact sites are often not available (in the case of unplanned events, for example) or are impractical (large-scale experiments), replicated control sites are usually possible. A common misconception is that the control site must be identical to the impacted site. Rather, controls only need be a random group of sites with the general features of physical characteristics, weather conditions, species, and the like. They should represent the dynamics of the conditions in which the impact has occurred (or is expected to occur). An impact affecting the abundance of the sampled population in the impacted site must cause the temporal pattern of abundance in that location to differ from the temporal pattern in the control locations (Figure 4.3). The selection of study sites is outlined later in this chapter.

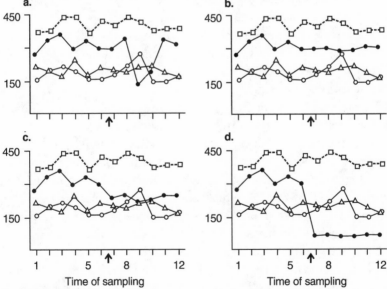

FIGURE 4.3.
Simulated environmental disturbances in one location (solid dots), with three controls—all sampled six times before and after the disturbance (at the time indicated by the arrow). (a–b) The impact is an alteration of temporal variance after the disturbance; temporal standard deviation times 5 in (a) and times 0.5 in (b). (c–d) A press reduction of abundance to 0.8 (c) and 0.2 (d) of the original mean. (From A. J. Underwood, "On Beyond BACI: Sampling Designs That Might Reliably Detect Environmental Disturbances," Figure 3. *Ecological Applications* 4:3–15. Copyright 1994.)

Desired Precision: Statistical Significance and Power

Data range from handwritten notes of personal observations to information obtained using sophisticated measuring devices. Whether data are in a qualitative form—such as "I observed that species A used oaks more than pine"—or of a more quantitative form—such as "I found a significantly ($P < 0.01$; t-test) greater use of oak ($x = 25.5\%$; SD = 5.20) than pine ($x = 11.0\%$; SD = 3.15)"—we have received information on an observed phenomenon. I suspect that today most of us would put more faith in the second example. But why should we? Given the information provided, there really is no reason to trust one observation over the other. If we were to add "On 75 occasions I observed . . . " to the first example and "DF = 6" to the second, I would certainly put my trust in the former (even though the latter does present a statistically significant difference). I make this conclusion because, without justifying one's sample size, no data set should be assumed to be valid regardless of the associated P value.

Finding a significant P in no way justifies a conclusion. Erroneous and often contradictory conclusions may be reached with variations in sample size, because alpha levels vary as sample size increases (Morrison 1984). In Figure 4.4, for example, note how increasing the sample size changes the estimates of the mean and standard deviation for the use of several plant species. Note too that the values are quite different depending on whether the data are collected using visual observations (observer's best guess) or instruments.

The probability of committing a Type I error (alpha: rejection of the null hypothesis when it is actually true) is inversely related to the probability of committing a Type II error (beta: failing to reject the null hypothesis when it is in fact false) for a given sample size. Lower probabilities of committing a Type I error are associated with higher probabilities of committing a Type II error. The only way to minimize both errors is to increase your sample size. Improving the power of a test has special importance to land managers. Looking at power another way, it tells us the likelihood that we will falsely reach a conclusion of no effect due to a treatment (or impact). If a test does not reject the null hypothesis of no difference but has low associated power, we are immediately warned that we just might be wrong. In such a case, caution is advised and further study is indicated.

Many formulas are available for estimating the sample size necessary to achieve a reliable result; virtually all basic statistics books provide them. (See, for example, Cochran 1977; Sokal and Rohlf 1981; Zar 1984; Petit et al. 1990.) An advantage of many of these methods is that they force you to indi-

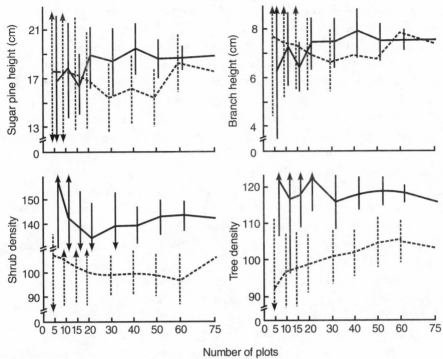

FIGURE 4.4.

Influence of sample size on the stability of estimates (dashed horizontal lines) and measurements (solid horizontal lines) of bird-habitat characteristics. Dashed or solid vertical lines represent 1 SD from point estimates for estimates and measurements, respectively. Variables shown are average height of sugar pine; average height to the first live branch of sugar pine; average number of shrubs within sample plots; and average number of trees within sample plots. (From W. M. Block et al., "On Measuring Bird Habitat: Influences of Observer Variability and Sample Size," Figure 2. *Condor* 89:241–251. Copyright 1987.)

cate the difference you want to achieve between your experimental units (to indicate treatment effect) and the alpha and beta you desire. This then leads to estimation of the sample size necessary to reach these goals. One difficulty is that most of the methods require you to have a good estimate of population variance. If this variance is unknown, it can be estimated from the literature or estimated as the study proceeds. Such *sequential sampling* allows you to refine your methods and effort as the study progresses and prevents you from simply gathering large volumes of data and hoping the effort will be adequate. Sequential sampling methods have been discussed by many workers (Kuno 1972; Green 1979; Block et al. 1987; Morrison 1988).

As indicated earlier, the number of samples you require depends on the precision of the answer you need. Green (1979) has shown that for a wide range of field data, the number of samples needed to reach a desired precision is independent of the unit of measurement (say, density) and is approximately equal to the inverse square of the desired precision. For example: if the density must be estimated with a precision such that 0.95 confidence limits are ±20 percent of the mean, then precision approximates 0.10 and the sample size approximates 100. If ±40 percent is adequate, then the sample size drops to about 25. (See Morrison 1988 for a review.) The number of replicates you need may also vary according to the statistical test planned. For example: in the randomized block design, a minimum of sixfold replication is necessary before significant ($P < 0.05$) differences can be demonstrated using the nonparametric Wilcoxon's signed-rank test, whereas only a fourfold replication is necessary using the Mann-Whitney U test (Hurlbert 1984).

Sampling from rare populations presents special problems. In such situations, it is impossible to gather a large, independent sample. Stratification of observations in time and in space, combined with a thorough analysis of the influence of sample size on results, will at least lend some confidence to your results. Further, various workers are developing statistical methods to address sampling from rare populations (Cochran 1977; Green and Young 1993; Thompson et al. 1998).

Johnson (1981) offers three guidelines for determining your sample size. First, examine the stability of the estimates, including both the mean and variance (sequential sample size analysis). Second, investigate the sources of variability and see how they compare in magnitude. Observer variability, temporal variability, variability in methods—all increase the variance in your estimates. Rather than simply increasing sample sizes to try to overcome this variability, it is more advantageous to reduce the variability from the start. And third, in the case of multivariate analyses the necessary sample size increases as variables are added to the equation. For each group in such an analysis, a bare minimum sample might be 20 observations, plus 3 to 5 additional samples for each variable.

Fancy statistical tests and complicated transformations of data cannot rescue a study that has an inadequate data set. Although we can all give reasons (make excuses) for failing to gather more data, there is no reason for failing to provide quantified justification for the samples we collected—and hence the impact of this sample on our conclusions.

In all ecological studies, we should be concerned with recognizing a meaningful *magnitude of biological effect*. That is, regardless of the statistical significance of a comparison, we must ask if this difference is meaningful biologically. Your restoration goals might be centered on achieving, say, a 20 percent increase in abundance of a breeding species. Your study must be designed to accurately identify this 20 percent increase and not just a "statistically significant" increase (which might be well below 20 percent). For recent articles on this issue for wildlife biologists see Johnson (1999) and Anderson et al. (2000).

Experimental Design

Experiments can be placed into two major classes: mensurative and manipulative. *Mensurative* studies involve measurements of uncontrolled events at one or more points in space and time. *Manipulative* studies, by contrast, always have two or more treatments with different experimental units receiving different treatments that may be randomly applied (Morrison et al. 2001:31–32). Neither type is necessarily more robust than the other; it all depends on the goal of your study. Manipulative experiments do have the advantage, however, of being able to test treatments and evaluate the response of something to varying degrees of perturbation—and to do so in a manner usually within your control. In developing an experiment it is essential to consider replication, controls, randomization, and independence. Hurlbert (1984) has presented an excellent review of the steps necessary for developing a strong experimental design; much of what follows in this section is based on his work.

Everyone is familiar with the general need for experimental controls, whether the study is mensurative or manipulative. Controls are required because biological systems vary over time. As noted by Hurlbert (1984), if we could be certain that a system was constant in its properties over time in the absence of an experimental impact, then separate controls would be unnecessary. (Pretreatment data from the site to be treated would be an adequate comparison.) In many experiments, moreover, controls allow you to separate the effects of different aspects of the experimental procedure.

Another definition of controls includes all of design features associated with an experiment. Controls *control* for temporal change and procedure effects. Randomization *controls* for (reduces) the experimenter's bias in assigning experimental units to treatments. Replication *controls* for the variability among replicates in a study. Interspersion (of study plots) *controls* for regular spatial variation in the experimental units. The term *control* may also refer to the homogeneity of the experimental units, to the precision of treatment procedures, or to the regulation of the physical environment in which the experiment is conducted. As Hurlbert points out, the adequacy of your experiment is based on your ability to "control" the physical conditions during the experiment and on your use of an adequate number of treatment controls (such as replicated control plots).

Replication reduces the effects of random variation (often referred to as noise) or error—thus increasing the precision of an estimate. Randomization reduces potential bias by the experimenter—thus increasing the accuracy of the estimate. Assume, for example, that the null hypothesis is that the abundance of a species does not differ between treated and control sites. If both areas are a mosaic of meadow and forest and the species in question occurs only in the meadow, then random sampling over the area (meadow and forest) would be an inefficient design that unnecessarily raises the error variation. This is true for two reasons: not only would the error include meadow versus forest differences but the ratio of variation between the control and treated areas to variation within them would be reduced—thus reducing the power of the test against the null hypothesis. Green (1979) offers further examples of how sampling effort should be allocated.

Independence refers to the probability of one event occurring that is not affected by whether or not another event has occurred. For statistical analyses, you should assume that the error terms are independently distributed. Departures from independence occur from correlations of the experimental samples in time and space (Sokal and Rohlf 1981). A somewhat common

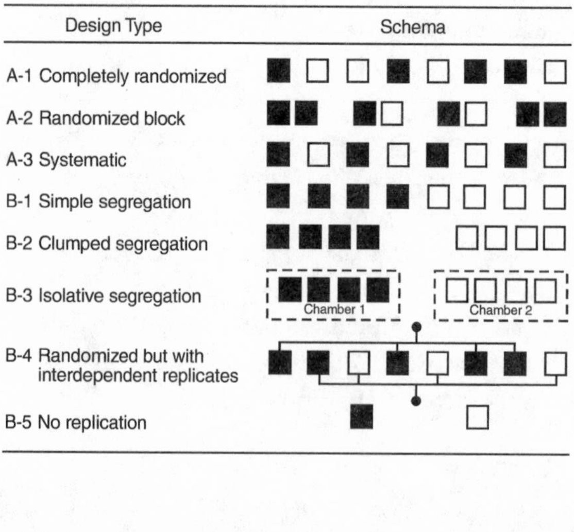

FIGURE 4.5.
Schematic representation of various acceptable modes (A-1 and so on) of interspersing the replicates (boxes) of two treatments (shaded, unshaded) and various ways (B-1 and so on) in which the principle of interspersion may be violated. (From S. H. Hurlbert, "Pseudoreplication and the Design of Ecological Field Experiments," Figure 1. *Ecological Monographs* 54:187–211. Copyright 1984.)

Design Type	Schema
A-1 Completely randomized	
A-2 Randomized block	
A-3 Systematic	
B-1 Simple segregation	
B-2 Clumped segregation	
B-3 Isolative segregation	Chamber 1 Chamber 2
B-4 Randomized but with interdependent replicates	
B-5 No replication	

misconception is that the application of nonparametric statistics relieves this assumption; it does not.

Hurlbert (1984) illustrates several ways in which treatments have been interspersed in a two-treatment experiment (Figure 4.5). The boxes in Figure 4.5 could represent any experimental unit (separate plots in a study area, separate study areas, individual animals). The "A" design types are considered acceptable; the "B" types are not. The following paragraphs summarize the main features of each design.

COMPLETELY RANDOMIZED DESIGN (A-1). This is the most basic method of assigning treatments. It is seldom used in field experiments, however, because it is usually difficult to achieve adequate spatial interspersion of treatments. This is because potential treatment plots are usually clumped in some fashion in nature (owing to animal or plant distribution, for example). Thus you run the risk (due to chance) of having treatments established in close proximity to each other (as in Figure 4.5).

RANDOMIZED BLOCK DESIGN (A-2). This method, commonly used in field studies, overcomes much of the problem outlined in the randomized design. And because treatments and controls are paired, it reduces variance due to differing environmental gradients between plots.

SYSTEMATIC DESIGN (A-3). This is another commonly used design in which you simply assign the first treatment randomly and then alter the remaining treatments and controls. A potential problem with this design in the field is that experimental plots might be spread along some environmental gradient such as elevation, soil, vegetation, and a host of other (often interrelated) factors. The block design (A-2) helps reduce the impact of such problems on test results.

SIMPLE, CLUMPED, AND ISOLATIVE SEGREGATION (B-1, B-2, B-3). These designs are seldom used in the field because they have the obvious disadvantage of segregating treatments from controls. (B-3 is primarily a laboratory design.) Thus certain differences that exist prior to the experiment might render test results spurious—and these problems may never be known to the experimenter. Differences in soil micronutrients because of local site conditions, for example, might be responsible for differences in plant response to some treatment, regardless of the treatment's effect. Moreover, any unplanned changes (such as catastrophes) after the experiment begins might only affect treatments (or only controls) because of their segregation.

PHYSICALLY INTERDEPENDENT REPLICATES (B-4). This design links experimental units to some common device such as a chemical source, heating element, or water source. Although this design may be used in any of the other designs, it is subject to spurious treatment effects. Moreover, it is subject to failure if the common device fails. Thus for both experimental design and practical considerations, this design should be avoided. Separate mechanical devices should be used for each experimental unit—which in effect eliminates this design. (It becomes one of the "A" designs.)

NO REPLICATES (B-5). Although the weaknesses of this design are obvious, it is nevertheless employed frequently. This is because ecological systems are often extremely large or by their very nature cannot be replicated within the area of interest. Few managers, for example, would be willing to replicate the single, unplanned field of exotic grass on their preserve simply for the sake of science. This problem is especially relevant with respect to environmental impact assessment (such as the accidental pollution of a stream by a toxic substance). Nonreplicated studies usually suffer from severe pseudoreplication (Hurlbert 1984). And even if they do not, their results cannot be generalized outside the immediate study area.

Lessons

The success of a restoration project depends on development of—and adherence to—a rigorous study design. Restorationists should avoid the temptation of rushing ahead without first developing all phases of the study, including likely statistical analyses. Even crude estimates of necessary sample sizes are essential if field efforts are to be maximized. There is no reason to undersample or oversample. Preliminary gathering of data, evaluation of the initial sampling efforts, and revision of the study's objectives and methods should be part of every study.

Although formal statistical tests remain a central part of many studies, they should be accompanied by a discussion of the magnitude of biological effect that was found. Simply finding a "statistically significant difference" means little when divorced from biological evaluation.

The statistical literature, and an increasing amount of the wildlife literature, provides in-depth guidance on designing, conducting, analyzing, and interpreting field studies. I urge all restorationists and wildlife scientists to increase their reliance on wildlife studies, especially during the design phase.

References

Anderson, D. R., K. P. Burnham, and W. L. Thompson. 2000. Null hypothesis testing: Problems, prevalence, and an alternative. *Journal of Wildlife Management* 64:912–923.

Bernstein, B. B., and J. Zalinski. 1983. An optimum sampling design and power tests for environmental biologists. *Journal of Environmental Management* 16:35–43.

Block, W. M., K. A. With, and M. L. Morrison. 1987. On measuring bird habitat: Influences of observer variability and sample size. *Condor* 89:241–251.

Cochran, W. G. 1977. *Sampling Techniques.* 3rd ed. New York: Wiley.

Gavin, T. A. 1989. What's wrong with the questions we ask in wildlife research? *Wildlife Society Bulletin* 17:345–350.

Green, R. H. 1979. *Sampling Designs and Statistical Methods for Environmental Biologists.* New York: Wiley.

Green, R. H., and R. C. Young. 1993. Sampling to detect rare species. *Ecological Applications* 3:351–356.

Hurlbert, S. H. 1984. Pseudoreplication and the design of ecological field experiments. *Ecological Monographs* 54:187–211.

Johnson, D. H. 1981. How to measure habitat—a statistical perspective. Pages 53–57 in D. E. Capen (ed.), *The Use of Multivariate Statistics in Studies of Wildlife Habitat.* General Technical Report RM-87. Washington, D.C.: USDA Forest Service.

———. 1999. The insignificance of statistical significance testing. *Journal of Wildlife Management* 63:763–772.

Kuno, E. 1972. Some notes on population estimation by sequential sampling. *Research in Population Ecology* 14:58–73.

Morrison, M. L. 1984. Influence of sample size on discriminant function analyses of habitat use by birds. *Journal of Field Ornithology* 55:330–335.

———. 1988. On sample sizes and reliable information. *Condor* 90:275–278.

———. 1997. Experimental design for plant removal and restoration. Pages 104–116 in J. O. Lucken and J. W. Thieret (eds.), *Assessment and Management of Plant Invasions.* New York: Springer-Verlag.

Morrison, M. L., and B. G. Marcot. 1994. An evaluation of resource inventory and monitoring programs used in national forest planning. *Environmental Management* 19:147–156.

Morrison, M. L., B. G. Marcot, and R. W. Mannan. 1998. *Wildlife-Habitat Relationships: Concepts and Applications.* 2nd ed. Madison: University of Wisconsin Press.

Morrison, M. L., W. M. Block, M. D. Strickland, and W. L. Kendall. 2001. *Wildlife Study Design.* New York: Springer-Verlag.

Osenberg, C. W., R. J. Schmitt, S. J. Holbrook, K. E. Abu-Saba, and A. R. Flegal. 1994. Detection of environmental impacts: Natural variability, effect size, and power analysis. *Ecological Applications* 4:16–30.

Peters, R. H. 1991. *A Critique for Ecology.* Cambridge: Cambridge University Press.

Petit, L. J., D. R. Petit, and K. G. Smith. 1990. Precision, confidence, and sample size in the quantification of avian foraging behavior. *Studies in Avian Biology* 13:193–198.

Romesburg, H. C. 1981. Wildlife science: Gaining reliable knowledge. *Journal of Wildlife Management* 45:293–313.

Sokal, R. R., and F. J. Rohlf. 1981. *Biometry.* New York: Freeman.

Stewart-Oaten, A., W. M. Murdoch, and K. R. Parker. 1986. Environmental impact assessment: "Pseudoreplication" in time? *Ecology* 67:929–940.

Thompson, W. L., G. C. White, and C. Gowan. 1998. *Monitoring Vertebrate Populations.* San Diego: Academic Press.

Underwood, A. J. 1994. On beyond BACI: Sampling designs that might reliably detect environmental disturbances. *Ecological Applications* 4:3–15.

Zar, J. H. 1984. *Biostatistical analysis.* 2nd ed. Englewood Cliffs, N.J.: Prentice-Hall.

CHAPTER 5

Fundamentals of Monitoring

A cornerstone of science is the acquisition of knowledge that improves our understanding of the natural world. Restorationists understand that their activities are improved through practice and experience. The better we document our activities, the better we will understand outcomes, both desired and undesired, and the easier it will be to convey this understanding to others. The way we accomplish these dual goals of understanding and conveying an outcome falls under the general rubric of "monitoring." Monitoring is a generally misunderstood process, however, as attested by the frequent misapplication of the term. Monitoring is a scientific endeavor and thus should be pursued with care. Although monitoring is sometimes treated as somehow different from research, or somehow requiring less rigor in design and application, this distinction is incorrect.

This chapter describes the process of monitoring as applied to restoration of wildlife and wildlife habitat. The principles discussed here apply, however, to most ecological issues. The parameter you select for monitoring in a restoration project will depend on the goal of your project. A project that seeks to enhance foraging resources, for example, would want to monitor the resource as well as its use by the species of interest. Moreover, the same project might want to determine whether providing that resource (say, high-protein plants) enhances reproductive success (and by how much). In Chapter 6 we will look at sampling methods that apply specifically to vertebrates.

Definitions

Monitoring is usually defined as "a repeated assessment of status of some quantity, attribute, or task within a defined area over a specified time period" (Thompson et al. 1998:3). Although the term has been used in a variety of contexts, here I will only use it in reference to approaches using repeated measurements taken at a specified frequency over time (Figure 5.1). Sometimes there is confusion between monitoring and the term *inventory* (and the related terms *baseline monitoring* and *assessment monitoring*). Although

FIGURE 5.1.
Tracking animals from the air (upper photo) is an efficient means of determining the distribution and habitat-use patterns of many species. Depicted here (lower photo) are woodland caribou in British Columbia. (Photos courtesy of Bruce G. Marcot.)

repeated samples may be taken to produce an estimate for a particular time period (month, season, year), an inventory is conducted by gathering data for a single time period. An inventory may be conducted to determine a point estimate for animal abundance during the winter, for example, or breeding success during the spring. This estimate tells us nothing about the change in these parameters over time. Repeating the inventory at another time period generally results in a monitoring study. Let me add, however, that you cannot have a rigorous and thus reliable monitoring program without a detailed study design. For other uses of the term *monitoring* see Thompson et al. (1998).

Thompson and colleagues (1998:6–7) have described a hierarchy of terms used to classify a monitoring program. An *element* is an item on which some type of measurement is made or some type of information is recorded. For example, an individual animal, a nest or den site, or a food item is an element. The next level of classification is the *sampling unit*, which is defined as a unique collection of elements. Note that by this definition, an element and the sampling unit can be the same thing—for example, a specific area or a specific nest. More often, however, multiple elements are included in a sampling unit, such as the animals that occur in a specific area or the number of nests in a field.

The complete list of sampling units is called the *sampling frame*. If the sampling unit is a study plot, for example, the sampling unit will contain a description (such as a listing or map) of all the sampling units. The sampling frame is, in essence, the physical way in which the elements are sampled. Here again there may be overlap in terminology: the elements can be sampled simply by drawing a boundary around an area without subdividing it into sampling units (see Figure 5.2).

Often we find it impossible to sample the elements in every sampling unit within the sampling frame. Perhaps the area under study is too large, for example, or certain sections of the area are inaccessible (owing to rough terrain or restricted access). Therefore, we take samples from the sampling frame; this sample is called the *sampled population*. The more thoroughly we take measurements from within the sampling frame, the more reliable our results will be.

The *target population* represents all the elements of interest: the target. Thus the target population represents all the elements within a defined time and space. If you are interested in determining the change in abundance of rabbits within a reserve for 3 years, for example, the target population is all the rabbits occurring in the reserve during the 3-year assessment period. To

a.

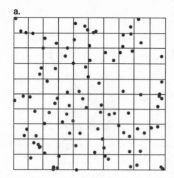

b.

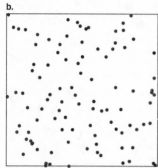

FIGURE 5.2.
(a) Example of a well-defined sampling frame composed of plots superimposed over randomly distributed elements (black dots). Although an area of interest is rarely so perfectly symmetrical, the concepts are the same regardless of the area's configuration (b). Example of an "undefined" sampling frame (represented only by a boundary with no plots delineated) superimposed over randomly distributed elements (black dots). (From W. L. Thompson et al., *Monitoring Vertebrate Populations*, Figures 1.3 and 1.4. Page 8. Copyright 1998, Academic Press.)

determine rabbit abundance, you must establish the sampling unit, sampling frame, and sampled population. Thus the target population is actually set before the study begins.

Inventory

We use an *inventory* to assess the status of resources. Inventories provide information on environmental characteristics such as the distribution, abundance, and composition of wildlife and wildlife habitats. Typically an inventory is confined to a specified area to give us a basic understanding of the wildlife present. Much of the environmental assessment done prior to development of an area, for example, is in the form of an inventory. The goal of such work is to generate a list of species occupying the area and perhaps gain a basic understanding of abundance and location. As we shall see in the next chapter, however, inventories are not quick-and-dirty surveys. It may take months of sampling to create an adequate species list (depending on the goal of the project).

Inventories associated with restoration projects are typically done to establish a set of baseline data on current conditions—that is, before initiating habitat-altering activities. Verifying a species' presence is relatively straightforward: if you detect the species by sight, sound, or sign, then you have established presence. But failure to detect a species does not necessarily mean it was absent when you sampled. (It might use the area during a different time period.) The sampling intensity and methods you use to determine a species' presence are critical if confidence is to be placed in your conclusions.

Monitoring

Monitoring studies assess changes in resources. Monitoring measures dynamics, therefore, whereas inventories index the state of the resource. The variables measured during monitoring often overlap—or might be identical—to those used in an inventory. Other variables, however, are unique to monitoring in that they measure rates such as survival and those that require repeated sampling (such as tree growth). Monitoring to determine trends must be conducted over a sufficient time period so that the population has been subjected to the usual range of environmental conditions (such as drought). Monitoring may include studies designed to evaluate the effects of a particular restoration treatment, or it may be used to estimate general trends without consideration for a particular activity. Monitoring studies often address objectives like these:

- Determine wildlife use of a resource (tree species, food type).
- Evaluate the effects of restoration on populations or a specific habitat.
- Determine changes in species diversity over large areas.
- Measure changes in population parameters (size, density, reproduction).
- Evaluate the success of predictive models.
- Assess floral and faunal changes over time.

Monitoring is sometimes placed in four overlapping categories. *Implementation monitoring* is used to assess whether a directed activity has been carried out as designed. Suppose a restoration project was designed to establish 40 percent willow cover at 3-m height within 4 years of implementation. Implementation monitoring would be done to evaluate whether this goal has been accomplished. *Effectiveness monitoring* is used to evaluate whether the stated action has met its objective. Suppose the reason for the 40 percent willow cover was to provide nesting locations for willow flycatchers. Effectiveness monitoring would determine whether the flycatcher was indeed nesting. *Validation monitoring* is used to determine whether management direction provides the right guidance to meet its stated objectives—for example, whether a mitigation plan actually resulted in recovery of a species in the area. *Compliance monitoring* is done when mandated by law or statute—for example, to determine whether the allowable take of an endangered species has been exceeded.

Sampling Considerations

Inventory and monitoring projects require an adequate sampling design to ensure accurate and precise measurements. Thus you will need to know the behavior, general distribution, biology, ecology, and abundance patterns of the resource of interest (Thompson et al. 1998). Gaining this knowledge usually calls for a thorough review of the literature, discussions with experts, and some preliminary field sampling to become familiar with the area.

The mobility of an animal population (within the restoration site, for example, or between the restoration site and surrounding locations) should also be considered when designing a monitoring project. That is, the decline of a species in the restoration area might be due to a regionwide decline in the population that has nothing to do with the restoration per se. Thus you need to consider the overall status (landscape perspective) of the species when designing—and especially interpreting—a monitoring effort. The issues raised in Chapter 1 have relevance here. Once these basic properties are understood, you can select a sampling methodology appropriate for the study's goals. Basic sampling issues were covered in Chapter 4.

Resource Measurements

Resources of interest can usually be measured directly or indirectly. If the project's goal is to increase the abundance of a specific resource, for example, one approach is to determine the density or abundance of that resource (via a bird count, for instance). In many cases, however, animal populations and other resources are difficult to sample because of their rarity or secretiveness. In such cases, investigators often rely on indirect measurements such as indices and indicator species.

Direct measures are variables that link clearly and directly to the question of interest. If you can obtain direct measurements, they are preferable to indirect measures. Direct measures establish a causal link between the variables being assessed. Assessing the abundance of prey being used by a foraging animal, for example, directly links the animal to a critical measure of its environment.

Nevertheless, *indirect measures* are widely used in inventory and monitoring studies. Such measures attempt to establish a surrogate for the direct link between variables of interest. In the preceding example, indirect measures of prey abundance might be sought because of the difficulty and expense involved in obtaining actual prey abundance. In this case, indirect measures might include counting tracks, burrows, feces, and other indications of the presence of the prey.

A broad class of indirect measures is known as *ecological indicators.* This concept was originally proposed by Clements (1920) to explain plant distribution based on specific environmental conditions, primarily soil and moisture. Wildlife species too may be tied to specific environmental conditions—which is, of course, the basis for describing a species' habitat (Block and Brennan 1993; Morrison et al. 1998). But actually quantifying the environmental requirements of animals on a species-by-species basis is extremely difficult. Many investigators, therefore, have tried to develop indicators—indirect measures—of the phenomenon of interest. Because of the mobility of most animals, relationships with the environment are not as strong for animals as they are for plants.

Care must be taken in selecting ecological indicators. You should be especially concerned with:

- Stating clearly what the indicator is indicating about the environment or resource
- Selecting indicators that are objective and quantitative
- Ensuring peer review for all monitoring programs using indicators (a standard that should apply to all study plans)
- Applying indicators at the appropriate spatial and temporal scale

The use of indicators is controversial and should be attempted only after careful evaluation. For guidance on this topic see Morrison (1986), Landres et al. (1988), and Morrison et al. (1998).

Habitat is often monitored as a surrogate for the animal. Monitoring a population to attain an acceptable statistical power can be very costly (Verner 1984), for example, so habitat is often monitored to index population trends for a species. We seldom have enough information for a species, however, to establish a strong link between habitat and population trend. As explained in Chapter 2, habitat is a complex concept that entails much more than the structure and floristics of vegetation.

Selection of Sampling Areas

Choosing the scale of resolution is perhaps the most important decision in a research project. This is because the spatial scale largely predetermines the procedures, observations, and results (Green 1979; Hurlbert 1984). A major step in a study is to clearly define the target population and the sampling frame. This first step establishes the universe from which samples can be drawn and the extent to which inferences can be extrapolated to areas outside your immediate study area. In many restoration projects, however, the physical size of the area is predetermined by management or regulatory issues. Relatively small areas (under 100 ha) are unlikely to support viable populations of many vertebrates unless the surrounding area is also suitable for habitation or at least provides passage routes. If the study's goal is to maintain population viability over the long term, the sampling area may include locations outside the immediate restoration site. To repeat: monitoring studies are simply a special type of research project and as such must have a rigorous study design (see Chapter 4).

The mobility of wildlife species may confound inferences drawn from the project area. In most ecological studies, a somewhat arbitrary decision is made to define the "population" potentially affected by a project. (See Chapter 1 for a detailed discussion of populations.) These decisions are usually based on information on habitat use, home range size, movement patterns, and the logistics involved in sampling the project area. Often the primary assumption is that animals within this investigator-defined area are the ones most likely to be affected by the project. Not measured, however, are the interacting effects that may have an impact on animals beyond the defined population area. Even though a great deal of thought might go into defining the sampling universe, the results of the monitoring may be questionable because factors outside the defined population may be affecting the animals within the area.

In some cases, the sampling universe is small enough that you can conduct a complete count (census) of the area. More often, however, the entire sampling universe cannot be surveyed—which means you must establish sampling plots. Here the primary considerations are: the shape and size of plots; the number of plots needed; and the placement of plots within the sampling universe. Determining shape and size is complicated by such factors as the methods used to collect data, biological edge effects, distribution of species under study (clumped versus random), biology of the species (nocturnal versus diurnal), and the logistics of data collection. Thompson et al.

(1998:44–48) have summarized the primary considerations and trade-offs in choosing a plot design. Long and narrow plots, for example, may allow for more precise estimates, but square plots have less edge effect. They conclude that no single design is optimal for all situations and suggest trying several designs in a pilot study (Chapter 4). Plot size depends largely on the biology and distribution of the species being studied. Large plots are needed for species with large home range sizes and for species with clumped distributions. Note that the species in Figure 5.3b will not be sampled adequately using the same plot size used for species in Figure 5.3a.

The number of sample plots and their distribution depend on several considerations including sample variance as well as the distribution and abundance of the species. Sample size should be defined by the number of plots necessary to yield precise estimates of the parameters of interest. Is the study's goal to identify a 10 percent decline in abundance with 90 percent confidence, for example, or is it to identify a 50 percent decline in abundance with 80 percent confidence? The sample size for the former will be substantially greater than that for the latter. Remember: you must establish

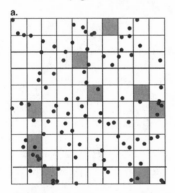

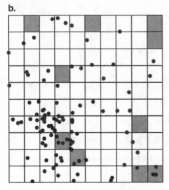

FIGURE 5.3.
(a) Simple random sample of 10 plots (gray squares) from a sampling frame with an underlying random distribution of individuals. (b) Simple random sample of 10 plots (gray squares) from a sampling frame containing a clumped distribution of individuals. (From W. L. Thompson et al., *Monitoring Vertebrate Populations*, Figures 4.5 and 4.6. Pages 137–138. Copyright 1998, Academic Press.)

specific effect sizes before you begin the monitoring work. Thompson et al. (1998) review many of these sampling issues.

Study Duration

The duration of a monitoring study depends on your study objectives, field methods, ecosystem processes, biology of the species under study, and logistical feasibility. A primary consideration is the temporal qualities of the ecological state or process being measured (plant succession, breeding cycles, generation time). These temporal qualities include the frequency, magnitude, and regularity of a process (Franklin 1989), which are influenced by biotic and abiotic factors alike. Thus you may find it necessary to collect data over a long time period to allow the population under study to face a reasonable range of environmental conditions (that is, to establish the reference condition). In many ecological systems, however, the major determinants of population fluctuations have a long periodicity. In piñon pine woodlands of the Great Basin, for example, major pine crops occur only every 6 to 10 years. Rodent populations in this region may fluctuate dramatically in response to this phenomenon (Morrison and Hall 1998). Four phenomena necessitate long-term studies:

- Slow processes such as plant succession and many vertebrate population cycles
- Rare events such as fire, flood, and disease
- Subtle processes where short-term variability exceeds the long-term trend
- Complex phenomena such as intricate ecological relationships

In reality, however, budget constraints often prevent long-term sampling. Therefore, innovative approaches are often called for when you need to achieve unbiased results from short-term monitoring. As the following paragraphs indicate, there are several alternatives to long-term studies (Strayer et al. 1986).

RETROSPECTIVE STUDIES. Retrospective studies have been used to address many of the same questions as long-term studies. Often they provide baseline data for comparison with modern observations. They can also be used to characterize slow processes and disturbance regimes and show how they may have influenced certain ecosystem attributes. Perhaps the greatest value of retrospective studies is in characterizing changes to vegetation and wildlife habitats over time. Dendrochronological studies yield information on the frequency and severity of historic disturbance events. This information can be used to reconstruct ranges of variation in vegetation structure and composition at diverse spatial scales. These studies can also be used to infer the short-term and long-term effects of different management practices on habitats, as well as the effects of disturbance regimes on habitats. Other potential tools for retrospective studies include databases from long-term ecological research sites, forest inventory databases, pollen studies, and sediment cores. With any of these studies, you must be aware of the underlying assumptions and limitations of the methodology. Dendrochronological methods, for example, often fail to account for the small trees that are consumed by fire and therefore not sampled. This limitation may result in a biased estimate of forest structure and misleading inferences about historic habitats.

SUBSTITUTION OF SPACE FOR TIME. You can substitute space for time by finding samples that represent the range of variation for the parameters of interest in order to infer long-term trends. Samples represent conditions for different points in time from some initiation event. For example, forests of various ages (seral stages) can be studied instead of waiting (a very long time) for a forest to grow. Morrison and Meslow (1984) studied the impact of herbicides on birds in clear-cuttings sprayed 2 years and 5 years earlier. The weakness of such studies is that true controls are unlikely to exist.

MODELING. Models are conceptualizations of how an ecological process might behave under various scenarios. Models can be as simple as verbal or pictorial descriptions of a process of interest, or they may be based on sophisticated mathematical constructs. For example, scientists use estimates of a population's demographic parameters (survival, reproductive output, recruitment) to project its size into the future. Habitat models link some measure of animal presence or abundance (presence/absence, low/high abundance, count index, density) with one (univariate) or multiple (multivariate) measurements of the animal's environment. Several techniques are especially popular: simple and multiple regression, discriminant analysis, and logistic regression. Morrison et al. (1994) used multiple regression, for example, to develop a relationship between animal abundance and vegetation measurements as a guide to restoration prescriptions (such as plant species composition and structural classes). Models of wildlife/habitat relationships may also be linked to models that predict changes through time in these key habitat characteristics to predict how populations might change under different restoration treatments. Verner et al. (1986), Patton (1992), and Morrison et al. (1998) present detailed descriptions of such models; see also Chapter 2.

Adaptive Management

As noted earlier in this chapter, inventory and monitoring provide information that can be used to evaluate and adjust management practices. Based on preliminary surveys, the project may be modified to enhance the outcome of the restoration or other management activity. Once the project has started, monitoring can then indicate whether its goals are being met. It is critical, of course, that you start with a rigorous baseline of data.

The concept of *adaptive management* or *adaptive resource management* (ARM) is centered primarily on monitoring the effects of land-use activities on key resources and then using the monitoring results as a basis for modifying those activities to achieve the project's goals (Walters 1986; Lancia et al. 1996). It is an iterative process whereby management practices are carefully planned, implemented, and monitored at predetermined intervals. Ideally the land management activities are implemented in stages so that desired outcomes can be monitored after each successive step. If outcomes are consistent with predictions, then the project continues as planned. If outcomes deviate from predictions, then management may proceed in one of three directions: it may continue, terminate, or change, depending on the project's goals.

Adaptive management is not a trial-and-error approach to restoration. Attempting to fix a problem after implementation is quite different from developing an action plan prior to the start of a project. This is where conceptual modeling assists—by allowing you to project the system under study into several likely outcomes based on the best available knowledge. Regardless of the specific approach, adaptive management offers a structure whereby clear goals are established and then monitored—and specific actions for responding to deviations are planned at the *outset* of the project.

Adaptive management can be summarized as a seven-step process that includes a series of feedback loops that depend largely on monitoring results (Figure 5.4). The primary feedback loops in Figure 5.4 are between steps 5–6–7, steps 2–7, and steps 7–1. The 5–6–7 loop is the shortest and usually the fastest to be initiated. It implies that management prescriptions are

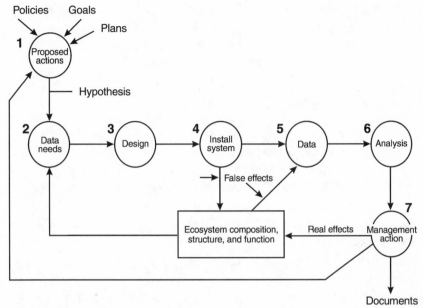

FIGURE 5.4.

A seven-step adaptive management system. False effects are data that reflect high instal-lation impacts, observer bias, location effects, and observer-caused disturbances during measurement. (From W. M. Block, personal communication.)

essentially working and need only slight adjustments if any at all. Probably the primary obstacle in this loop is the time lag between project implemen-tation and the point where the project has advanced sufficiently to monitor for effectiveness (Moir and Block 2001)—for example, allowing tree growth to proceed to the point where survival and structural development can be assessed. Because this time frame may be prolonged, numerous factors may complicate management's ability or willingness to complete monitoring and implement needed changes (Morrison and Marcot 1995). Consequently, the loop is often severed and feedback is never provided. The second loop (2–7) indicates that monitoring was poorly designed, the wrong variables were measured, or monitoring was badly executed. Regardless of the exact reason, monitoring failed to provide reliable information to permit informed deci-sions on the efficacy of past management or future management directions. The 7–1 feedback loop is the one typically associated with adaptive man-agement: this is when a decision must be made regarding the course of future management and monitoring activities. If monitoring was done correctly, informed decisions can be made for future management directions. If monitoring was not conducted or was done poorly, however, another opportunity to provide a scientific basis for resource management was lost. Unfortunately, the latter is the rule rather than the exception.

Thresholds and Trigger Points

In designing inventory and monitoring studies, you must establish some benchmark that signals a need for subsequent action. For adaptive management to be effective, threshold levels of key resources must be established prior to starting the project. These thresholds may include specific stocking densities of plants, foliage cover, or even the presence or other measure of animal activity. Violating such a threshold—either high or low—will then trigger a specific management action. Thus thresholds and triggers are a central component of adaptive management.

Triggers lead to specific responses—perhaps modification or even termination of an activity. Suppose, for example, your restoration plan calls for attaining an overstory of 30 percent cottonwood cover and an understory of 80 percent shrub cover—along with cowbird density not to exceed 1 female/ha and an absence of feral cats—within 5 years of treatment. If these thresholds are not achieved, then specific actions are prescribed at the start of the study. Here planting (or thinning) of trees, as well as trapping and removal of cowbirds and cats, would be likely prescriptions. Now suppose your monitoring plan follows standard protocols. Five years after restoration monitoring, you determine that too many large trees are present and cowbird abundance is 3 females/ha. Thus two of the thresholds have been exceeded and you take remedial actions to return the system to the desired condition. Most likely you would have planned 10-year, 15-year, and further thresholds as well. This is a very different scenario than waiting until year 5 to determine that something is amiss and then starting hurried discussions on how to fix the problem. More important: without establishing thresholds beforehand, there is no way to design an appropriate monitoring program. The intensity required to determine that no cats were present, for example, is much more rigorous than for simply determining that fewer than five cats were present. In addition, there may have been no reason to even attempt certain restoration activities in this hypothetical area if political considerations had prohibited cat or cowbird control as an option.

Lessons

My goals in this chapter were to give you a general sense of the topics to consider and the effort needed in designing inventory and monitoring projects. We learn by conducting field studies and evaluating the results of our restoration efforts. Thus we must be clear about what determines a project's success. It is essential, therefore, that we implement well-conceived inventory and, especially, monitoring programs. Inventory and monitoring projects are research and must be conducted with rigor. Because many of our projects do not allow for adequate replication—and because we often must proceed without adequate preliminary data—adaptive management presents a viable method for tracking a project's results and making wise modifications. Adaptive management is not, however, synonymous with trial and error.

In the following chapter I review many sampling protocols specific to terrestrial vertebrates. To ensure the success of an inventory or monitoring project, great care must be taken to link your project's goals with a rigorous sampling strategy.

References

Block, W. M., and L. A. Brennan. 1993. The habitat concept in ornithology: Theory and applications. *Current Ornithology* 11:35–91.

Clements, F. E. 1920. *Plant Indicators.* Washington, D.C.: Carnegie Institution.

Franklin, J. 1989. Importance and justification of long-term studies in ecology. Pages 3–19 in G. E. Likens (ed.), *Long-Term Studies in Ecology: Approaches and Alternatives.* New York: Springer-Verlag.

Green, R. H. 1979. *Sampling Designs and Statistical Methods for Environmental Biologists.* New York: Wiley.

Hurlbert, S. H. 1984. Pseudoreplication and the design of ecological field experiments. *Ecological Monographs* 54:187–211.

Lancia, R. A., et al. 1996. ARM! For the future: Adaptive resource management in the wildlife profession. *Wildlife Society Bulletin* 24:436–442.

Landres, P. B., J. Verner, and J. W. Thomas. 1988. Ecological uses of vertebrate indicator species: A critique. *Conservation Biology* 2:316–328.

Moir, W. H., and W. M. Block. 2001. Adaptive management on public lands in the United States: Commitment or rhetoric? *Environmental Management* 28:141–148.

Morrison, M. L. 1986. Birds as indicators of environmental change. *Current Ornithology* 3:429–451.

Morrison, M. L., and E. C. Meslow. 1984. Response of avian communities to herbicide-induced vegetation changes. *Journal of Wildlife Management* 48:14–22.

Morrison, M. L., T. Tennant, and T. A. Scott. 1994. Laying the foundation for a comprehensive program of restoration for wildlife habitat in a riparian floodplain. *Environmental Management* 18:939–955.

Morrison, M. L., and B. G. Marcot. 1995. An evaluation of resource inventory and monitoring programs used in national forest planning. *Environmental Management* 19:147–156.

Morrison, M. L., and L. S. Hall. 1998. Responses of mice to fluctuating habitat quality. I: Patterns from a long-term observational study. *Southwestern Naturalist* 43:123–136.

Morrison, M. L., B. G. Marcot, and R. W. Mannan. 1998. *Wildlife-Habitat Relationships: Concepts and Applications.* 2nd ed. Madison: University of Wisconsin Press.

Patton, D. R. 1992. *Wildlife Habitat Relationships in Forested Ecosystems.* Portland, Ore.: Timber Press.

Strayer, D., J. S. Glitzenstein, C. G. Jones, J. Kolasa, G. E. Likens, M. J. McDonnell, G. G. Parker, and S.T.A. Pickett. 1986. *Long-Term Ecological Studies: An Illustrated Account of Their Design, Operation, and Importance to Ecology.* Occasional Paper 2. Washington, D.C.: Institute of Ecosystem Studies.

Thompson, W. L., G. C. White, and C. Gowan. 1998. *Monitoring Vertebrate Populations.* San Diego: Academic Press.

Verner, J. 1984. The guild concept applied to management of bird populations. *Environmental Management* 8:1–14.

Verner, J., M. L. Morrison, and C. J. Ralph. 1986. *Wildlife 2000: Modeling Habitats of Terrestrial Vertebrates.* Madison: University of Wisconsin Press.

Walters, C. 1986. *Adaptive Management of Renewable Resources.* New York: MacMillan Publishing.

CHAPTER 6

Sampling Methods

The previous chapter outlined the basic concepts of inventory and monitoring, including study design and statistical analysis. The foundation for a useful monitoring study is, of course, its design. But regardless of the design, a study will fail to produce reliable data if the methods you use to assess the distribution and abundance of the target animal are inappropriate.

There are numerous methods of sampling animals—most of them designed to sample a particular aspect of a population. The sampling method you choose must be applicable to the study's goal. *Point counts* are not superior to *spot mapping* to assess birds, for example, and *area-constrained surveys* are not superior to *pitfall traps* for assessing herpetofauna. Each method has strengths and weaknesses that must be matched with your project's goals. Although the scientific literature provides guidance on the applicability of a particular method to specific project goals, the literature is not much help when it comes to the frequency and especially the intensity of application necessary to achieve reliable results—largely because the project's goals drive sampling intensity. Very few papers address these topics, however, relying instead on untested assumptions regarding the application of commonly used methods.

This chapter focuses on the proper application of sampling methods. Rarely will one method sample all species in a vertebrate class adequately. Thus each description of a method summarizes its strengths and weaknesses, the sampling intensity necessary to achieve different levels of confidence regarding the results, and suggested refinements of standard techniques and potential ways of combining methods to improve a survey.

Principles

This section outlines the general sampling principles that apply to all techniques. It bears repeating: "study the study design" to ensure that your sampling intensity is adequate to address your project's goals. This rule is important for all studies, but it is especially critical for long-term studies. Pilot studies are seldom included as an integral part of project design. Thus the first few months of most projects become "pilot studies" by default because of the errors made and lessons learned in trying to implement the design. It is better, of course, to design the pilot study beforehand so that you have control over these errors and lessons. Chapter 4 presents a detailed discussion of study design.

Observer Bias and Training

We all take numerous biases to the design of a study: our training predisposes us to prefer certain sampling methods; we might believe that animals respond in certain ways to environmental conditions, which tends to narrow our sampling focus; time and funding constraints limit our ability to sample in the way we would prefer; and so forth. Not only do biases influence the way we design a study, but additional biases are inserted into field sampling when multiple observers are used to gather data. Ignoring observer-based variability may lead to incorrect conclusions because of artifacts, spurious relations, or irreproducible trends (Gotfryd and Hansell 1985:224). Here I want to concentrate on the reduction of bias in field sampling.

Intraobserver reliability is a measure of an observer's ability to obtain the same data when measuring the same behavior on different occasions (Martin and Bateson 1993:32–34)—in other words, the ability of an observer to be precise in his or her measurements. *Precision* describes the repeatability of a measurement and is not synonymous with accuracy. Because we seldom know the actual behavioral pattern—that is, the "true" pattern—we cannot measure an observer's accuracy directly. Assessing an observer's reliability in field-based studies is especially difficult because animals seldom repeat their behavior in exactly the same fashion. One test is to videotape animals and then repeatedly present individual sequences to the observer in some random fashion. Researchers can use the results of such trials to estimate the degree of observer reliability.

Interobserver reliability measures the ability of two or more observers to obtain the same results on the same occasion (Martin and Bateson 1993:117). To what extent is interobserver reliability a problem in field studies? Ford et al. (1990), for example, found that comparisons of foraging behavior of individuals of the same species in different areas or different years, recorded by different observers, had to be treated cautiously. Problems were particularly evident when observers had not agreed on a standard method of observation or classification of terms beforehand. Differences in observers' experience accounted for much of the variability noted.

Researchers conducting behavioral studies should be aware that their presence is likely to influence an animal's activities. Wild animals are constantly vigilant for predators and competitors, and your presence is likely to heighten their awareness. Such heightened awareness is termed *sensitization*. Further, the animal probably knew you were there long before you ever saw it. The waning of responsiveness, termed *habituation*, is usually considered to be a form of learning (Immelmann and Beer 1989). Animal species vary widely in their ability to learn. Research has shown that birds and mammals have the ability to perform both temporal and numerical operations in parallel (Roberts and Mitchell 1994). Many corvids can remember the location of food caches for months, for example, and can recall which caches they visited previously. Rosenthal (1976) has presented a detailed analysis of the effects of the researcher in behavioral studies.

Animals that seem habituated to an observer's presence have in fact adopted a modified pattern of behavior that allows them to keep the observer under surveillance. Animals adjust their behavior according to the costs and benefits associated with different courses of action (hiding versus fleeing, for example). Moreover, detection of a potential predator (or human observer) may precede the observable response by a significant period of time. Roberts and Evans (1993) found that when sanderlings (*Calidris alba*) were approached by a human, they minimized both the number of flights they made and the distance of each flight.

Gotfryd and Hansell (1985) used four observers to independently sample eight plots in an oak-maple forest near Toronto, Canada. They found that observers differed significantly on 18 of the 20 vegetation variables measured. (Their study addressed only the precision of estimates between observers; no measure of the accuracy of their results was conducted.) Block et al. (1987) used several univariate and multivariate analyses to test for differences among three observers in estimating plant structure and floristics. They found that visual estimates by the three observers differed significantly for 31 of the 49 variables they measured. Perhaps the most confounding aspect of using multiple observers was the unpredictable nature of variation among observers. And when samples from different observers are pooled, sampling bias can increase. Ganey and Block (1994) used three observers to sample plots for canopy closure using two different estimation techniques (spherical densiometer and sighting tube). They found significant variation among observers in estimates of canopy cover for both methods. Results were more consistent for the sighting tube, however.

The biases associated with estimations of animal abundance emerge in habitat studies as well. This is because many of our analytic procedures correlate animal numbers with features of the environment. Clearly a study that has low bias among habitat characteristics may be ruined by biased count data (and vice versa). Dodd and Murphy (1995), for example, evaluated the accuracy and precision of nine techniques used to count heron nests. Although they found rather high error rates among the techniques, observer bias was low for most methods. Interestingly, the highest observer bias was found for their point counting technique—a result apparently due to the varying choices made by observers of the optimum vantage point from which to count nests at colonies.

You can reduce interobserver variability by following a set of well-defined criteria for selecting and training observers. Although designed for bird censusing, the steps outlined by Kepler and Scott (1981) for bird counting methods apply to most types of sampling:

- Screen applicants carefully to eliminate obvious visual, aural, and psychological factors that increase observer variability.
- Organize a rigorous training program to reduce inherent variation.
- Institute periodic training sessions to counter observer "drift" and thus recalibrate their recording to fixed standards.

In a field experiment, Scott et al. (1981) found that trained observers could estimate the distance of a singing bird within 10 to 15 percent of the true distance. Observers' reliability also increases when they come to understand why certain behavior is categorized in a certain way. Each type of behavior should be carefully defined in writing. *Probe*, for example, means "insert bill beneath surface of substrate." In protracted studies, definitions and criteria tend to drift with the passage of time as observers become more familiar with animal behavior and possibly lax in their evaluations. Careful and repeated training will help solve this problem. Further, efforts have been made to standardize terminology in various disciplines. Remsen and Robinson (1990), for example, developed a standardized vocabulary for avian foraging studies. The methodology of Schleidt et al. (1984) for a standard ethogram is a useful starting point for behavioral studies.

Plant ecologists have long recognized differences among data collection techniques. (See Cooper 1957; Lindsey et al. 1958; Schultz et al. 1961; Cook and Stubbendieck 1986; Hatton et al. 1986; Ludwig and Reynolds 1988.) Although the cost of measuring plant structure and floristics in an adequate number of plots is usually great in terms of both time and money, the ramifications of not following a rigorous sampling design are severe. Again, it is better to limit the scope of a study so that the data you do collect are gathered properly; preliminary sampling and analysis will help you to spot potential problems.

Types of Information

Often the most basic information of interest is a list of the species present in an area over a specific period of time, usually a season (breeding, winter). The number of species present is termed *species richness*. Studies of species richness are often conducted to evaluate poorly known sites or to gather basic information for a detailed study. Rigorous techniques (transects, point counts, and the like) are almost always needed to gather a complete species list. Often a survey begins by reviewing the literature and distribution maps and by conducting walking ("birding," mammal tracking) surveys throughout the study area, spending as much time as needed to identify each species present, and perhaps gathering a qualitative estimate of abundance ("common," for example, or "rare").

Distribution studies specify where animals—usually one or a few species of interest—occur and do not occur. In most bird guides the distribution maps represent an accumulation of casual records, perhaps formal surveys, and the general distribution of the appropriate environmental conditions. Recently states have initiated formal surveys of breeding birds, usually on a county-by-county basis. Coverage of the area of interest should be uniform or needs to be measured and reported. This allows users to interpret the results properly regarding species distribution. Proper interpretation is especially important when the data are being used to evaluate the presence of legally protected and other species of concern. With the exception of game mammals, such systematic surveys are not available for vertebrate groups in most areas.

Population monitoring monitors trends in abundance or other demographic parameters over time. Because population numbers fluctuate owing to weather, food availability, disease, catastrophes, and many other causes, it is difficult to separate these interacting factors from human influences. The ability to separate the multitude of factors shaping population trends is a cornerstone of monitoring. Substantial variation in population numbers from year to year may mask a long-term trend, for example, thus delaying the implementation of remedial action. Another cornerstone of population monitoring is the use of consistent methods over time. Repeatable counts need not be accurate in the sense that they represent the absolute number present, but they must use the same technique and intensity of application.

Assessments of habitat are used to predict the impact of land-use practices and the distribution and abundance of animals over time and space. Two basic techniques are used. One technique is to describe the habitat of a species (from the literature, through a specific study) and identify it in the field using a variety of methods (aerial photographs, ground surveys). The other technique is to find statistical relationships between the number of animals present (or simple presence/absence) and associated environmental features. Because of the high variation in environmental conditions in most areas, a large sample of study plots is usually needed. And in many cases we are most interested in the rare species, which means large samples will be difficult to obtain. (See Chapter 2 for a detailed presentation of habitat assessment.)

Sampling Errors

The true value of the phenomenon we are trying to determine—density, species richness, habitat—is usually unknown to us. The difference between the true value and our estimate is termed *error*. Error is, in turn, composed of *normal variation* and *bias*. A study with low variation is considered to have high precision, and one with low bias is considered to have high accuracy. Precision and bias (inaccuracy) may vary independently within a single study. Precision measures how close our estimates are to one another, regardless of how closely they approximate the true value. A tight cluster of darts that widely miss the target, for example, has high precision but low accuracy. The results of a study may be biased because of inadequate effort, differences in vegetation and other environmental features (see Figure 6.1), animal behavior (noisy versus secretive species), and numerous other factors.

Precision can be measured. It can also be improved by increasing the sample size (see Figure 6.2). But as a general rule, precision increases only in pro-

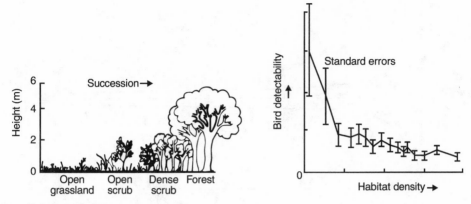

FIGURE 6.1.
Birds are more conspicuous in open areas than in dense woodland. The hypothetical species is equally abundant across the succession but might appear more abundant in the grassland and young trees where it is more easily detected. This effect is particularly serious if the bias arises from the same source as the object of study (such as the effect of forest succession on bird communities). (From C. J. Bibby et al., *Bird Census Techniques*, Box 2.7. Copyright 1992. Reprinted by permission of Academic Press.)

portion to the square root of sample size: to double the precision obtained from 10 samples would take 30 samples; doubling it again would take another 120 samples. Here again we see why the success of a monitoring study depends on determining the necessary precision before you initiate field sampling. The variability inherent among samples interacts with the goals of a study to determine the precision you need. That is: if natural variation and bias are low, you may need fewer samples than when natural variation and bias are high. Again it bears repeating: pilot studies and sample size analyses are critical to the success of a monitoring study. The common sources of bias in animal counting—the observer, the method, the sampling effort, weather, and so forth—have been discussed for herpetofauna (Heyer et al. 1994), for birds (Ralph and Scott 1981; Bibby et al. 1992; Ralph et al. 1995), and for mammals (Wilson et al. 1996).

General Considerations

Questions concerning animal diversity fall into two general categories: those related to vegetation types or specific areas and those related to certain species or groups of species. The primary goal of the first category is to determine the species that occur in specific vegetation types or specific areas. The second type, species-based studies, may focus on one or more populations

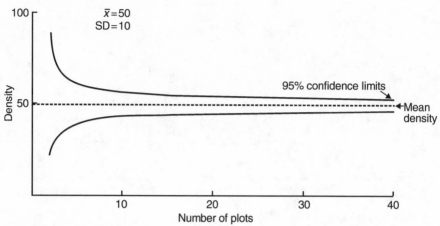

FIGURE 6.2.

Relationship between number of plots sampled and our confidence in the result for a hypothetical population. (From C. J. Bibby et al., *Bird Census Techniques*, Box 2.2. Copyright 1992. Reprinted by permission of Academic Press.)

across space or time to determine, for example, geographic or ecological distribution. Although species-based studies may be conducted as an inventory, usually you will need a battery of techniques to give you a thorough description of the species. The techniques you select and the intensity of their implementation depend, of course, on your project's goals. Species-specific surveys are driven by the behavior of the species of interest. Certain techniques, no matter how popular in the literature, may be unsuitable for determining even the presence of various species. The following sections outline the techniques available for inventorying and monitoring vertebrates.

Amphibians and Reptiles

Heyer et al. (1994) present a detailed description of most of the sampling techniques available for amphibians and reptiles (herpetofauna) and have summarized the 10 most commonly used techniques (Table 6.1). Note that the techniques vary in the type of information gained as well as the cost in terms of sampling time, money, and labor. Obtaining a density estimate (numbers per unit area) and a complete survey of the species present is relatively expensive in terms of time. Obtaining estimates of species richness and density may take weeks or months of sampling. Obtaining a complete inventory of species in an area takes an extended period of time—and the other techniques listed with "low" or "medium" time are no substitute for the

TABLE 6.1. Standard Techniques Used to Sample Amphibians and Reptiles, Type of Information Gained, and Relative Time and Cost

Technique	Information gained[a]	Time[b]	Cost[c]	Personnel[d]
1. Complete species inventories	Species richness	High	Low	Low
2. Visual encounter surveys	Relative abundance	Low	Low	Low
3. Audio strip transects	Relative abundance	Med	Med	Low
4. Quadrat sampling	Density	High	Low	Med
5. Transect sampling	Density	High	Low	Med
6. Patch sampling	Density	High	Low	Med
7. Straight-line drift fences and pitfall traps	Relative abundance	High	High	High
8. Breeding site surveys	Relative abundance	Med	Low	Med
9. Breeding site drift fences	Relative abundance	High	High	High
10. Quantitative sampling of amphibian larvae	Density/relative abundance[e]	Med	Med	Med

Source: W. R. Heyer et al., *Measuring and Monitoring Biological Diversity: Standard Methods for Amphibians*, Table 4. Page 77. Copyright 1994.

[a]Designations are hierarchical: techniques that provide density estimates also provide relative abundance and species richness. But if a technique provides only relative abundance, an additional technique must be used to provide density.

[b]Relative time investment.

[c]Relative financial cost: high = relatively expensive; medium = moderately expensive; low = relatively inexpensive.

[d]Personnel requirements: high = more than one person required; medium = one or more persons recommended; low = can be done by one person.

[e]Some methods included in technique 10 give relative abundance only, and some provide density estimates.

complete inventory. Hence there is no easy way to obtain a thorough species survey. The 10 standard techniques, it should be noted, are not mutually exclusive. Surveys at breeding sites (ponds, creeks), for example, are often used to supplement general techniques such as visual encounter surveys and pitfall trapping.

Scott (1994) has detailed the techniques for generating a species list for an area. Typically these methods involve identifying the herpetofauna in all possible microsites, day and night, by searching the surface of the substrate and turning over rocks, logs, and other cover. Such techniques have been used for both short-term and long-term monitoring. Although these methods can be used for sampling many of the species likely to be present in an area, secretive, canopy-dwelling, fossorial, and deep-water species often require specialized techniques (Heyer et al. 1994).

Often it takes years to sample herpetofaunas with numerous species or secretive species. Short-term sampling cannot give you much insight into the total number of species present at a site. Thus in many restoration studies—where the project site is small—long-term, intensive sampling is necessary. The results of short-term sampling depend on a host of variables including weather (before and during sampling), experience of observers, sampling effort at different microsites, and number of different techniques used.

Visual encounter surveys (VESs) are those in which observers walk through an area for a prescribed period of time systematically searching for animals. Time is usually expressed as the number of person-hours. The VES is considered appropriate for both inventory and monitoring studies, although the caveats regarding secretive species apply here as well. The VES is used to determine the species richness of an area and to estimate relative abundance of species. VES may also be done in a plot (technique 4 in Table 6.1), along a transect (technique 5), or in a user-defined patch of vegetation (technique 6). The goals and study design of such techniques are somewhat different from many VES studies in that the plots and transects are usually placed in some random fashion that is appropriate for a statistical comparison of sites, habitats, or other environmental features. Crump and Scott (1994) have detailed the implementation of various designs and derivations of VESs. Jaeger and Inger (1994) and Jaeger (1994a, 1994b) have described the related techniques of quadrat, patch, and transect sampling, respectively.

The *time-constrained survey* (yielding the number of individuals of different species collected per person-hour) is basically a VES that is restricted to some predetermined time period such as 4 person-hours. Scott (1994) considers the time-constrained survey a less robust form of the VES because of this time limitation.

Audio strip transects (technique 3) are used to count calling animals (usually frogs). (See Zimmerman 1994 for details.) The width of the transect depends on the detection distance of the various species' calls. The counts are then used to estimate:

- Relative abundance of calling individuals
- Relative abundance of all adults
- Species composition
- Breeding or microsite use
- Breeding phenology

This technique is basically an adaptation of singing bird surveys, which have well-developed methodologies and analytic techniques.

Pitfall traps (with or without drift fences) are commonly used to sample herpetofauna (Figure 6.3). These methods yield estimates of species richness and relative abundance and are particularly effective in capturing secretive, fossorial species. A pitfall trap is a container placed in the ground so that its open end is flush with the surface. Animals are captured when they fall into the trap. Pitfall traps are constructed from small cans, plastic buckets, or PVC pipe and are 40–50 cm deep and 20–40 cm wide (Jones et al. 1996). Corn (1994) presents a detailed description of trap design and placement and explains basic survey methods. (See also the discussion under "Mammals.")

Funnel traps, another valuable means of assessing species richness, are extremely useful for sampling snakes. Funnel traps consist of rounded tubes or rectangles (window screen or hardware cloth) with an inwardly directed funnel-shaped opening at one or both ends. Snakes are directed into the funnel by the drift fence but find it difficult to negotiate their way

FIGURE 6.3.
Pitfall traps established to monitor various small vertebrates in chaparral vegetation in southern California. Upper photo: researcher examining a pitfall constructed of a plastic bucket. Lower photo: drift fence extending out from a pitfall trap. (Photos courtesy Zoological Society of San Diego.)

back out of the trap. During 1040 trap-days in northern California, for example, Swaim (unpublished data, 1996) captured 18 species of small vertebrates in funnel traps in oak-bay woodland—including 3 species of amphibians, 3 species of lizards, 8 species of snakes, and 4 species of small mammals.

Artificial wooden cover has been used to sample herpetofauna (see Fellers and Drost 1994 for details). This technique is relatively new, however, and has not been extensively tested. It has several advantages: the number of cover boards can be standardized for comparisons between sites and environmental conditions; there is little variability from observer to observer, especially compared with time- or area-constrained techniques; it avoids disturbing natural cover (such as logs); and it is inexpensive and requires little training. There are, however, several disadvantages: the technique provides only an index of population abundance; not all species use artificial cover; the use of artificial cover may decline as temperatures increase and the environment becomes dry; and it is difficult to place cover boards in many vegetation types (shrubs, tall grass). Fellers and Drost (1994) summarized successful use of this technique. There are few situations, however, in which cover boards alone would give you a thorough assessment of the herpetofauna in an area.

Night driving is a form of nonrandom line transect in which the transect is a paved road. Often the technique is used to sample herpetofauna that are attracted to the heat radiating from the road in the evening as the air cools. The technique alone, however, cannot give you reliable quantitative estimates of absolute abundance for most species or relative abundance of species for an area. Driving shortly after dawn may locate animals killed by passing cars the previous evening. Through road driving, for example, Morrison and Hall (1999) located new species for the Inyo and White mountains of eastern California that were not recorded during three seasons of intensive pitfall traps and VES sampling.

Birds

There are numerous reasons for counting birds. In this section I outline the many techniques for doing so. For detailed works critically reviewing sampling methods see Ralph and Scott (1981), Bibby et al. (1992), and Ralph et al. (1993). Much of this section is based on the excellent coverage of counting techniques given by Bibby et al. (1992).

The sampling technique you choose should match the spatial scale of your study. Are the results intended to apply to a large geographic area or a small plot? Moreover, the type of data you need will determine both the method you choose and the intensity of application. Are simple presence/absence data sufficient, for example, or is density (number per unit area) necessary? And what magnitude of error is acceptable? As noted in Chapter 4, a properly designed study begins with a clear statement of goals. Bibby et al. (1992) present a detailed explanation of the standard sampling techniques, including sample data forms, recording methods, and methods of interpreting data. The following sections highlight the key components of these standard techniques.

Territory or Spot Mapping

The conspicuous behavior of many bird species, especially vocal passerines, is a cornerstone of the territory or spot mapping technique. Singing males display from various locations in order to attract and breed with a female, thus establishing an area usually defended from conspecific males: the territory. The spot mapping method uses this behavior to establish rough territorial boundaries and obtain an estimate of the density and location of birds in the study area. These data can also be used to correlate changes in the number and size of territories with environmental conditions (see Chapter 2).

This technique is the most time consuming of the standard bird counting methods. Thus it is usually applied to small areas (under 20 ha) in studies where detailed information is needed on the location of birds, territory size, and habitat use. Studies of rare and threatened species often incorporate this method as part of an assessment of nesting, demographics, and behavior. Essentially an observer crosses the study area repeatedly, recording the specific location and behavior of each bird (of the species of interest) seen on a scale map of the area. (See Bibby et al. 1992:boxes 3.4 and 3.5 for details.) Usually the observer makes 10 visits to the study area.

Bibby et al. (1992:65) list the following points to consider when spot mapping:

- Mapped counts take considerable time to complete in the field and analyze.
- Their best feature is that, unlike other techniques, they result in a map of the distribution of birds.
- Mapping has a good chance of obtaining an estimate of absolute abundance (especially when combined with color marking).

- There are established rules for field recording and data analysis that make it possible to compare across years and studies.
- You will need to establish guidelines for analyzing maps.

Line Transects

Line transects are a commonly used technique for assessing the distribution and abundance of birds over large areas (more than 20 ha) of relatively uniform terrain. To avoid double counting of individual birds, transects need to be widely spaced (usually more than 200 m). Since calculation of density is usually unnecessary, it is not discussed herein.

Detecting birds while walking requires excellent birding skills. Thus this technique is sensitive to subjective bias from the observer. It is also susceptible to bias from factors affecting the detectability of birds—bias that must be controlled when possible and considered in the interpretation of data. Although transects can be used year round, bird detectability changes substantially between seasons due to bird behavior (especially changes in vocalization), weather, and foliage cover. Bibby et al. (1992:67) think transects may be more accurate than point counts because violations of assumptions regarding distances between birds and observer have an impact that rises linearly for transects and by square for point counts. In practice, however, point counts are actually a specialized application of transects, and each technique has a preferred use.

Bibby et al. (1992:71) have depicted various field designs for transects (see Figure 6.4). The specific form you use depends on the goals of your study and the structure of the vegetation in the study area. If you need abundance indices in open vegetation (grassland, low shrub, marsh), for example, then either no distance measuring or recording within a fixed belt should suffice (Figure 6.4a or b1 or b2). But if you have to calculate density estimates, you will need more specific measurements of distance (Figure 6.4c or d).

Usually transects are not used to develop fine-scale assessments of bird habitat use. This is because you need an adequate number of bird detections along the transect to be able to relate to vegetation and other environmental conditions—which usually requires at least 100 m of transect. Thus transects are best suited for large-scale assessments of bird abundance and habitat relationships. Although a transect can be subdivided (into 50-m segments, for example), this defeats the purpose of the technique. Point counts

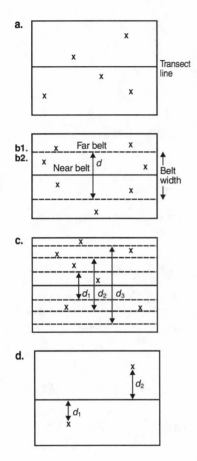

FIGURE 6.4.
Field designs for transects. (a) No distance measuring; all birds are counted. Although this method is simple, different species are counted on different scales because of differing detectabilities. Five birds (x) have been recorded. (b1) Fixed belt: all birds are counted within a predetermined fixed belt (near belt). This lowers the total count but removes distant records of the more conspicuous species. In this case, four birds have been recorded and three birds have not. (b2) Two belts: all birds are counted but attributed to one of two belts. This is an effective method that is simple to use in the field. Relative densities can be estimated. Four birds have been recorded in the near belt and three in the far belt. (c) Several belts: birds are attributed to one of several belts of fixed width (d_1–d_3). This is harder to do in the field because distances have to be estimated to greater precision. It is often more satisfactory to use the other methods cited here. Counts in the first four belts were 1, 3, 2, 1. (d) Distances are measured to all birds. Distances are perpendicular to the route even if the bird was ahead when detected. This is the hardest method to use in the field but generates the best data for estimation of densities. Birds were recorded at distances d_1 and d_2. (From C. J. Bibby et al., *Bird Census Techniques*, Box 4.3. Copyright 1992. Reprinted by permission of Academic Press.)

provide a more useful method of developing fine-scale bird/habitat relationships.

Often it is necessary to determine distances because of substantial differences in detectability between species (Figure 6.5). Although this does not present a problem when you are comparing a species within a single vegetation type, it complicates comparing a species across several vegetation types. As noted earlier, there are analytic techniques for calculating density or indices of abundance when detectability changes across space or time.

Bibby et al. (1992:80–81) have summarized the assumptions associated with transect counts and list several points to consider when designing and implementing this technique:

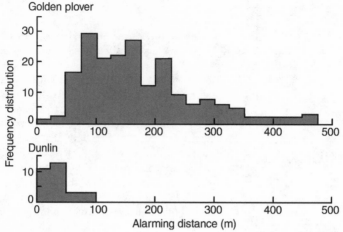

FIGURE 6.5.
Numbers of golden plover and dunlin detected at different distances from the observer.
Golden plovers are noisy and conspicuous, especially as they react to an observer and
give alarm at greater distances than do dunlin. Dunlin are cryptic and sit tight and
thus are not detected beyond 100 m. This difference between species must be taken
into consideration when designing breeding-site sampling methods. (From C. J. Bibby
et al., *Bird Census Techniques*, Box 4.4. Copyright 1992. Reprinted by permission of
Academic Press.)

- Transects are particularly suitable in extensive, open, uniform, or species-poor areas.
- Transect counts require a high level of bird identification skills.
- Transect counts are the most efficient of all standard techniques in terms of data gathering per unit of effort.
- Transects generate less detail than mapping.
- Habitat measurements can be made along the transect.

Although there are no standard rules for recording birds on transect counts,
consideration should be given to selection and location of routes, number of
visits (three visits is a minimum standard), walking speed, distance estimation, and observer and other bias.

Point Counts

Point counts are similar in conception to transect counts. In fact, a point
count may be viewed as a transect of zero length conducted at zero walking
speed. Point counts were developed for application in rough terrain where
simultaneously walking and observing birds was difficult or dangerous

(Reynolds et al. 1980). This technique has gained widespread acceptance in all but open terrain. It has also been frequently applied in studies of bird/habitat relationships.

Point counts are similar to transects in that a high level of observer skill is required. Because points are used in relatively closed vegetation, most birds are recorded based on vocalizations rather than sight. By waiting at each point, the observer has more time per unit area to identify birds. And because observers are stationary, there is no noise associated with walking along a transect.

Distance estimations can be applied to point counts in a fashion analogous to transects. Thus the guidelines regarding no distance estimation, a fixed distance (such as a 50-m-radius plot), or individual distance estimations apply here. Because studies have shown that the detectability of many breeding passerines begins to decline rapidly at 50 to 75 m from the observer, a distance of 50 m is often chosen to establish a fixed plot. Using fixed plot radii is usually unacceptable, however, when you need to compare results between different vegetation types.

Bibby et al. (1992:86) say that point counts are efficient in that many points can be counted per unit time relative to other techniques, including transects. For example: if counts can be conducted on 40 days during the breeding season, three counts must be conducted per point (a minimum standard), and if 10 points can be visited in a morning (a typical scenario), then about 133 points could be established. This does not address the issue of independence between points, however. Avian ecologists typically spread points just far enough apart to avoid double-counting birds at adjacent points—a typical distance is 200 to 300 m between points. These points are usually sampling the same environmental conditions within a small area, however, and thus should probably not be counted as independent samples. In Figure 6.6, for example, panel (a) establishes one transect, whereas panel (b) establishes six points. Independence of the points would depend on the scale of the map. Regardless of the issue of independence of points, this technique is preferred over transects in developing fine-scale assessments of bird/habitat relationships. Habitat parameters are sampled near the point (or within the fixed circular plot) and related to the birds counted there. Ralph et al. (1995) discuss point counting applications in detail. For a summary of the methods used to develop wildlife/habitat relationships, including habitat assessment techniques, see Chapter 2.

Bibby et al. (1992:104) cite the following points to consider in implementing a point count:

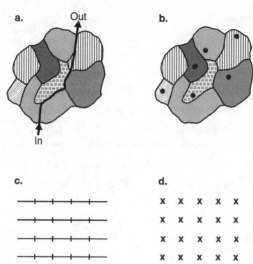

FIGURE 6.6.

(a) In a fine-grained environment such as a wood, a transect following an access route might not be very representative. It would not be easy to divide the bird records into vegetation types. Indeed, in this example two of the vegetation types have not been sampled at all. (b) In the same place, point counts could be set out at random or systematically to represent the full range of conditions present in the wood. Each point could also have the conditions around it recorded. (c) In open country, transects could be set out in a way to cover more of the ground and divided into sections for recording birds and habitats. (d) The equivalent design for point counts would theoretically record fewer birds but would take about the same time to execute. If birds were flushed ahead of the observer, however, as is generally the case in open ground, this would be a poor design because the observer walking up to a point would scare all the birds away. (From C. J. Bibby et al., *Bird Census Techniques*, Box 5.1. Copyright 1992. Reprinted by permission of Academic Press.)

- Point counts are suitable for monitoring birds in woody, shrubby, and rough terrain.
- Point counts are suitable for large areas but do not yield the detailed information that is available through mapping.
- Point counts are more efficient in terms of information collected per unit of effort than mapping, but they may be less efficient than transects (depending on the issue of independence of points).
- Point counts are especially useful in developing bird/habitat relationships.

There are no fixed rules for point counting, but careful thought needs to be given to selection and location of points, number of visits (three is a minimum standard), duration of each visit (5 to 7 minutes is typical), measuring of distance to birds, and observer bias.

Marking Birds

Marking birds is a specialized technique that can be used to estimate population size, quantify habitat use, determine survival rates, measure movements, and assess various behavioral activities. The advantage of having individually marked birds is clear: you are able to quantify the behavior and outcomes (survival, productivity) for known individuals rather than obtaining a sample of individuals of unknown identity. Thus you can analyze birds by sex and age, known lineage, known place of birth, and so forth. Marks include metal and plastic leg bands, patagial (wing) tags, neck collars, plumage dyes, radio transmitters, and a variety of other specialized techniques. For most studies, however, capture and marking are unnecessary.

This technique requires specialized training and licensing and is fairly labor-intensive when particular species are being studied. Bibby et al. (1992:105–106) note the following considerations involved in a capture and marking study:

• Can you catch enough individuals to obtain worthwhile results?
• Will the marking harm the bird or influence its behavior?
• How difficult will it be to recapture the birds if needed?
• Will you be able to readily observe the marks in the field?

Bibby et al. (1992), White and Garrott (1990), and Bookhout (1994) present detailed descriptions of capture and marking techniques as well as statistical means of analyzing mark-recapture (or resighting) data.

There is some confusion over the use of mist-netting as a technique for monitoring birds. Mist-netting is a standard means of capturing birds so they can be fitted with a U.S. Fish and Wildlife Service aluminum band. These bands have unique numbers and are used primarily to determine migration routes and movements of passerines. There are numerous banding stations throughout North America, many of them operated by amateur ornithologists. A large-scale program of netting and banding termed Monitoring Avian Productivity and Survivorship (MAPS) seeks to determine trends in bird survival and productivity. This program is appropriate for determining population trends over a large geographic area. General netting is seldom a viable method of population monitoring on small areas (under 1000 ha), however, because birds move widely in searching for food, mates, and shelter. Moreover, most bird populations are composed of a large proportion of adult, nonbreeding individuals. Termed "floaters" because of their widespread movement relative to breeding adults, these birds are usually indistinguishable from breeders in appearance and sometimes behavior (male floaters often sing, for example), and thus make it hard to interpret capture results. It is especially difficult on small areas to determine where the birds you captured came from. In small areas, moreover, netting alone cannot be used to determine age ratios (in order to calculate productivity) because young of the year disperse widely from their natal area. Thus you must use caution when planning a study that incorporates netting. Of course, many of these weaknesses also apply to counting methods.

Comparison of Counting Methods

Bibby et al. (1992:box 5.2) offer a useful comparison of mapping, transect, and point counts applied to the same hypothetical woodland. As we have noted throughout this book, your decision on which technique to use rests primarily on the goal of your study. Bibby and colleagues also offer recommendations for individual species and groups of species (colonial nesting birds, raptors, nocturnal birds) that require specialized techniques or adaptations of standard techniques.

Mammals

Numerous techniques are available for sampling mammals and assessing their diversity in a study area. Mammals present special sampling difficulties, however, because even a relatively small area may contain species that exhibit substantially different ecologies. Throughout temperate North America, for example, plots of only 1 or 2 ha may harbor several fossorial species, shrews, several species of *Peromyscus*, small carnivores (weasel, skunk), and a bat roost—and, moreover, medium-sized and larger species likely traverse the area on a daily basis. This diversity presents a special challenge to the observer.

Wilson et al. (1996) present a detailed description of most mammalian sampling techniques. Indeed much of the material summarized here is based on their review. The numerous field techniques for studying species richness and abundance of mammals fall into two categories: observational—either direct observation or indirect observation of sign—and capture techniques.

Voucher Specimens

Small and medium-sized species are often difficult to identify in the field. This is because identification may involve subtle morphological characteristics, including skull and other bone characteristics. Further, immatures are often extremely difficult to identify in hand. Thus *voucher specimens* must be taken so that an archival record of the study can be maintained. Vouchers physically and permanently document data by confirming the identity of mammals used in a study and assuring that the study can be repeated, reviewed, and reassessed accurately (Reynolds et al. 1996).

Observation of Nonflying Mammals

Various observational techniques are available to determine a mammal's presence and abundance in a study area. These techniques involve detection of animals by sight, sound, and physical evidence (notably tracks and dens). Here I review several of the standard sampling techniques. Rudran et al. (1996) describe specialized techniques such as aerial and night surveys.

DRIVE COUNTS. You can obtain an estimate of total abundance in a small area with a *drive count*. In this technique, an area is completely surrounded by people who count animals as they are forced to leave the area. Naturally, this method is suitable only for species that are highly visible and cannot hide within the study area. Drives have been used most successfully for ungulates. Rudran et al. (1996) detail specific techniques and analyses for drive counts.

TRANSECTS. Line and strip transects are conducted as for other vertebrates. (See the earlier discussion on herpetofauna and birds.) Further, the theoretical assumptions are the same as for other vertebrates. In the *line transect* technique, an observer travels along a line recording the animal itself, its call, or its sign. The *strip transect* differs from the line transect in assuming that all animals (or animal sign) within the strip are seen. For sign the strip has traditionally been narrow—such as the width of the observer's outstretched arms (*arm-length transect*). Tracks, burrows, and scat are often counted in such a manner. For observations of individual animals, the width of the strip is determined by the visibility of the animal and the denseness of the vegetation.

Road counts have been used in many studies of mammals, especially ungulates (including nocturnal spotlighting surveys). The technique is similar to either the line or strip transect except that the road itself is the transect. Some people favor this technique because travel through the study area is easy, and large areas can be covered rapidly. Road surveys are naturally biased, however, because roads tend to follow land contours and are not established randomly.

Roads may be useful for obtaining a long-term index of animal abundance, but results should not be extrapolated to the population as a whole.

QUADRAT SAMPLING. *Quadrat sampling* (square or rectangular plots) is similar to strip transects in that all animals or their sign within the sampling plot are assumed to be observed. In practice, however, quadrat sampling usually involves small, randomly (or stratified randomly) placed plots (1 to 100 m^2) that are surveyed intensively for sign (burrows, tracks, scat). This technique is useful for obtaining an index of animal activity over time.

Observation of Bats

Historically bats have been viewed by the public as carriers of disease. In many areas, they are considered pests and exterminated when they inhabit buildings and other locations in proximity to humans. Although bats—like many other mammals—may serve as disease vectors, they seldom come into contact with humans and, moreover, consume large quantities of insects that are harmful to crops. In fact, many farmers are now establishing roosting boxes to encourage bats to inhabit their lands. Thus there is increasing interest in bat ecology, including a better understanding of their interactions with humans. But because of their past—and, in many areas, continuing—status as pests, we have little information on the distribution and abundance of bats. Most field guides simply provide general descriptions of bat distribution and habitat use. Like birds, many bats migrate between breeding and wintering areas. During migration, bats require roosting and feeding sites. Thus a thorough sampling of bats must be conducted during all seasons of the year. Bat sampling for monitoring studies is usually concentrated near roosts. Basic sampling methods include direct roost counts, disturbance counts, nightly emergence counts, hibernaculum counts, and ultrasonic sound detectors. Kunz et al. (1996) and Jones et al. (1996) detail general sampling methods and specific capture techniques.

ROOST COUNTS. *Direct roost counts* are done by one or more observers who systematically count all visible bats in a diurnal roost (Figure 6.7). For bats that roost by hanging from horizontal surfaces (from the ceiling of a building or cave, for example), this method can obtain a nearly total count. Many bats, however, roost in crevices or behind other structures that obscure an observer's view. Direct counts may be used to monitor bat abundance within a specific roost that is small and has a simple physical structure, but they are unsuitable for comparing numbers between roosts because of uncertainties about within-roost movement of bats. (The number of bats that cannot be seen is likely to change between surveys.)

Disturbance counts overcome the problem of bats roosting out of sight or moving between counts. In this technique, bats are stimulated to take flight during the day and are counted as they become airborne. Once airborne, bats may be visually counted or photographed as they emerge from the roost. The success of this technique depends on the observer's skill and the sensitivity of the bats to disturbance. Bats will be miscounted if some fail to fly,

FIGURE 6.7.
Upper photo: Abandoned mines, such as the one shown here in Inyo County, California, are often used as roosting and maternity locations by bats. Lower photo: Townsend's big-eared bats using an abandoned mine, Inyo County, California. (Photos by the author.)

do not leave the roost, or continually leave and reenter the roost. This method should only be used during favorable weather, moreover, and should never be used during breeding because of disturbance to pregnant and nursing females.

Nightly emergence counts are conducted as bats leave their diurnal roosts. (Kunz et al. 1996 have described a related technique termed *nightly dispersal counts* that follows the methodology described here.) This is an effective means of counting bats that roost in inaccessible locations or in places with a complex internal structure—and an effective way of minimizing disturbance to the roost. It is difficult, of course, to actually count bats as they emerge. Except in small roosts with a single opening, you will usually need several observers. For use in monitoring, you will have to standardize observer effort, observer skill, positioning of observers, weather conditions, and time of observation. Still photography and videography can sometimes be used to record emerging bats. In some situations, night-vision devices may enhance observations.

Hibernaculum counts are done in midwinter to quantify hibernating bats. Extreme caution must be used, however, for disturbance to hibernating bats may lead to their death. Because of the importance of adequate hibernaculum to the maintenance of bat populations, these roosts must be located and protected (Szewczak et al. 1998; Kuenzi et al. 1999).

ULTRASONIC DETECTORS. *Ultrasonic sound detectors* are a useful technique for noting the presence of bats and can often be used to identify species. These devices are used primarily to assess species richness at nonroosting sights during the night, such as water holes and riparian foraging areas. Distinguishing echolocation calls of different species is based on species-specific features such as frequency composition, changes in frequency with time, and duration of pulse retention. Most calls are highly structured; frequencies range from 20 to 200 kHz. Several detailed reviews of the equipment—which is constantly being improved and lowered in cost—are available. (See, for example, the review by Kunz et al. 1996.) Bat detectors have several limitations, however, with regard to monitoring studies. Above all, they require a high degree of skill in recording and interpreting data. We are only beginning to study how bat calls vary within and between species and geographic locations. Moreover, reference calls—obtained by recording the calls of known species from a study site—are often necessary. Current research indicates that the calls of some species (such as *Myotis*) overlap extensively, making species identification suspect. Further, the echolocations of some species can only be detected at close range—meaning that detectors cannot always assess species richness fully. And some bat species are seldom captured—which means you will need a combination of techniques to fully gauge species richness.

Capture Techniques

Several methods of capturing mammals have gained general acceptance in the literature. But simply applying methods that have been used in other studies does not guarantee they are appropriate to your situation. In fact, few of these "standard" methods have been validated for adequacy. (Chapter 4 covers study design and, in particular, preliminary sampling.) Frequently used methods do offer an appropriate starting point, however, from which you can make modifications as you evaluate preliminary data in relation to your project's goals. Each study must consider the following issues:

- The type of capture device
- The type and quantity of bait
- How the traps are distributed (the trapping array)
- How long the traps will remain open (the trapping interval)
- How animals will be processed (handling, identification, marking)

Jones et al. (1996) have outlined the generally accepted techniques for capturing mammals. Here I summarize the most frequently used methods.

DEVICES. There are several commonly used devices for capturing small mammals (less than 150 g). Finding the most appropriate device depends on the species involved and the purpose of your study (Figure 6.8).

Snap traps are used for killing rodents. These traps are useful for rapid assessment of species richness, especially in remote areas that are hard to reach. They are also the device of choice when your goal is to estimate abundance through the removal method (see Lancia and Bishir 1996). Snap traps should not be used, of course, when protected species are suspected to be present. Although small mousetraps are available in most hardware stores, they are not powerful enough to kill anything instantly but the smallest rodents and should not be used. The most effective snap traps are Museum Special mouse and rat traps (Woodstream Corporation, Lititz, PA 17543).

Box traps are the most commonly used device for live-trapping small

FIGURE 6.8.
Capturing, handling, and marking small mammals is a common means of quantifying changes in species composition and animal health as part of monitoring studies. (Photo by the author.)

mammals. The most popular traps are manufactured by H. B. Sherman Traps, Inc. (Tallahassee, FL 32316). Although many people refer to "Shermans" in the same way that tissue paper and photocopies are described by popular brand names, there are several manufacturers of small mammal traps. (See Wilson et al. 1996:app. 9 for a comprehensive listing.) Box traps are used primarily for studies in which animals are not killed and the goal is to derive an index of abundance (along with data on species richness). These traps are available in several sizes so they can be effective with animals of various dimensions and weight. Because box traps are usually left unattended overnight, specific steps should be taken to protect the animals:

- Provide adequate bait.
- Provide insulative material (shredded paper or fiber batting of polyester, cotton, or wool).
- Partially bury the trap and cover it with grass and a small amount of soil.
- Cover trap with foliage or a wooden cover (such as a cedar shingle).
- Never place a trap unprotected on the soil surface.

Protecting traps in this way also reduces trap disturbance by predators. Fieldworkers have an ethical obligation to ensure that animal discomfort is minimized. Trap deaths will occur, however, and can be used as a source of voucher specimens. Box traps are also used for medium-sized (up to 5 kg) and large (over 5 kg) mammals. Such traps are heavy, however, and hard to position in adequate numbers in the field. Thus padded leghold traps are often used instead of box traps. Legholds must be checked frequently to prevent injury, stress, and predation to the animal. Because of public concern, several states (Arizona, California, and others) have recently restricted the use of leghold traps.

Pitfall traps are an effective means for sampling very small mammals (under 10 g), especially shrews (*Sorex*). The design and placement of pitfalls for mammals is the same as described earlier for herpetofauna (including the use of drift fences). In fact, most pitfalls capture both mammals and herpetofauna regardless of the project's objectives. Pitfalls must remain open for an extended period of time in order to determine species richness and abundance—often for 2 or 3 months. Traditionally, pitfalls were used as kill traps by filling them with liquid (often a mixture of water and antifreeze to prevent evaporation) because of the difficulty of checking them every day. Although pitfalls can be run essentially the same as live traps if they are left dry and provided with insulation, batting, and bait, in some locations they will collect water. (Drain holes can be added in dry environments, but they cannot be used in moist areas.)

Jones et al. (1996) have summarized specialized capture techniques that can be used for medium-sized and especially large mammals, including nets, dart guns, and drugged bait.

BAIT. A trap's effectiveness is usually enhanced by providing bait. Moreover, the food supplies energy for animals during their confinement. Virtually nothing is known, however, about how different species react to different types of bait. The standard rodent bait is rolled oats (packaged oatmeal) and peanut butter or a commercial bird seed mix. Different baits can be used when several traps are used at one location. About one tablespoon of bait is usually sufficient. Certain species appear to be attracted to specific baits—for example, the recommend bait for the salt marsh harvest mouse (*Reithrodontomys raviventris*) is ground walnuts and bird seed. For carnivores, pieces of flesh (mammal remains, fish, poultry) and scents (urine, rotten eggs, fish oil) can be used as attractants. Be sure to provide water in box traps being used for medium-sized and large mammals. And take care to ensure that your bait does not interfere with the trap mechanism (by lodging, for example, under the treadle in box traps).

TRAPPING ARRAYS. The same type of trap and bait are used for both inventories and estimation of abundance. But the placement of traps—the trapping array—depends on the goal of your project. For inventories that seek to obtain species richness, you can place traps along transects. The transect's length is determined by your project's goals and the nature of the environment (patchy versus homogeneous, for example). As a starting point, Jones et al. (1996) recommend a transect at least 150 m long with traps spaced every 10 to 15 m for small mammals. For larger species, the transect length should be based on the animal's home range size. In medium-sized species, a trap spacing of at least 100 m is recommended. Because of the wide variation in movements of larger species, however, standard distances cannot be fixed. Parallel transects can be used to sample larger areas and become, in essence, large-scale trapping grids. Transects, if repeated over time, can be used to create an index of abundance.

As Jones et al. (1996) point out, transects cannot be used to generate a density estimate. This is because, by definition, density is the number of animals per unit area. In practice, you should not try to estimate density (except perhaps in single-species studies when you need to know density, as for an endangered species analysis). Typical rodent trapping grids range from 8 × 8 (64 traps) to 10 × 10 (100 traps) with a single trap at each grid point and 10 to 15 m spacing between traps. Jones et al. (1996) recommend 10 × 10 arrays with two traps at each grid point. Although this is probably a good recommendation, at least initially, I do not think this type of grid is adequate for assessing density. First, density appears to be an artifact of sampling area (grid area) in many mammals. For example, Smallwood and Schonewald (1996) have shown that most of the variation in density estimates for large carnivores was due to the size of the study area chosen by the researcher. Second, recent radiotelemetry studies have shown that rodents appear to have a much larger home range than that determined through trapping grids. In fact, using a grid of a predetermined size to determine home range is an exercise in circular reasoning. And third, indices of abundance are adequate surrogates for density if you use an adequate sampling effort (Figure 6.9). For most monitoring purposes, then, an index of abundance is adequate if the

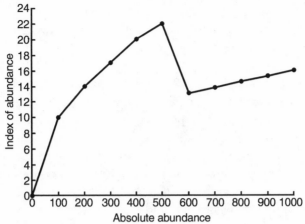

FIGURE 6.9.

Relationship between indices of abundance and absolute abundance. (From M. J. Conroy, *Measuring and Monitoring Biological Diversity: Standard Methods for Mammals*, Figure 38. Page 181. Copyright 1996, Smithsonian Institution Press.)

sampling effort is standardized. Grids of traps can also be used for larger species (but remember the caveat on trap spacing).

TIME INTERVAL. Most monitoring studies aim to determine species richness and abundance. Because restoration efforts usually expect to see an increase in richness and abundance over time, you must obtain reliable estimates of these parameters for evaluating the project's progress. Standardizing the time that traps are open (the number of consecutive days) is not appropriate in most situations. This is because it usually takes a longer trapping period to quantify species richness and abundance in species-rich areas. Few studies report the relationship between trapping effort and the fauna obtained. Most rodent studies use a 3-day trapping session; a few go for 4 or 5 days. In fact, Jones et al. (1996) recommend that you operate traps on a schedule that coincides with five or six activity periods for the species of interest. For most small mammals, this means a 5- or 6-day trapping period. Jones and colleagues recommended at least 7 days for medium-sized mammals. I agree—and suggest that you use preliminary trapping to determine the interval. The longer the interval, of course, the more difficult it becomes to sample multiple study areas. But obtaining inadequate samples from many sites is not productive.

A cautionary note: many small mammals will be captured repeatedly within a trapping session. If your traps are not adequately provisioned with insulative material and bait, these animals gradually lose weight and often die toward the end of the session.

ANIMAL HANDLING. Jones et al. (1996) offer a useful description of how to remove small mammals from traps and handle them during measuring, marking, and other tasks. Field personnel must also become acquainted with the diseases that may be carried by small mammals—rabies, plague, and hantavirus, for example. Moreover, all field personnel are subject to tick-borne Lyme disease. Project directors should contact the Centers for Disease Control and Prevention (CDC) and local health agencies to discuss such concerns in the study area and find out the latest methods of prevention.

TECHNIQUES FOR BATS. Bats are captured for a variety of reasons—to determine species identification, sex-age ratios, and reproductive status, for example, and to apply marks. Jones et al. (1996) have summarized trapping methods, and Kunz (1988) has detailed the methodology. Here I comment briefly on several popular techniques and add a few cautions.

Several factors should be considered when you are designing an inventory project that involves capturing bats: weather conditions, moonlight, daily activity patterns, seasonal behavior (migration, dispersal, breeding), vegetative structure (short versus tall canopy), colony size, and the type, number, and arrangement of capture devices. Such factors determine both the species composition and the number of bats captured. Thus it is essential that these factors be standardized or otherwise accounted for when you are comparing study sites through time. Extreme caution must be used when trapping near roosts because of the problem of capturing too many bats and the time it takes to remove and process each one.

Because bats tend to concentrate their foraging activities in areas of high insect abundance—near water sources and over riparian trees, for example— most trapping has been done near such sites, which biases our understanding of bat activity. Bats do forage away from wet areas, however. For small project areas without any apparent attractants for bats, try using the observation techniques described earlier to determine whether trapping is necessary.

Mist nets, often the same type as those used for birds, are the most commonly used method to capture bats. Although traps are usually placed in arrays around sites frequented by bats, such as roosts (caves, buildings)

FIGURE 6.10.
Example of mist-net placement near water.

and water holes, they can be placed in any location thought to be frequented by bats (Figure 6.10). Results are biased, of course, by the height
that the nets are placed. For bats, as for birds, nets can be raised on poles
or a pulley system to capture high-flying individuals foraging along the top
of vegetation (Jones et al. 1996:fig. 19). Nets placed in the subcanopy may
attain capture rates up to 10 times higher than those placed at ground
level.

Harp traps are another popular capture device (Figure 6.11). These consist of a rectangular frame crossed by a series of vertical monofilament lines.
When a bat hits the wires, it falls into a bag beneath the trap from which it
can be easily removed. Often these traps are used to capture bats in confined
locations (narrow streams, cave entrances), as opposed to the large sampling
area usually covered by mist nets. Harp traps and mist nets are not necessarily equally effective in capturing bats. Harp traps, for example, may be more
effective in capturing large bats (over 150 g).

Lessons

Animal ecologists have spent considerable effort developing rigorous sampling methods for vertebrates. Often I am asked about the best sampling
method for a species. This question can only be answered in the context of

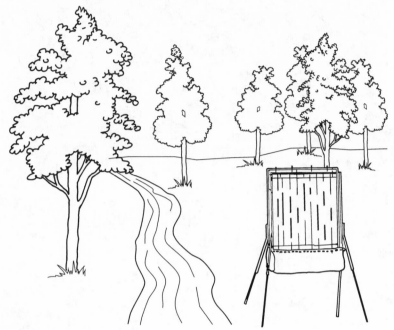

FIGURE 6.11.
Example of harp-net placement.

a study's goals. That is, your sampling methods and sampling intensity must match the goals or your project. As noted in Chapter 4, there is no justification for oversampling or undersampling. And as we learned in this chapter, restorationists will find a well-developed literature on vertebrate sampling methods to help them design their projects. Do not hesitate to be creative and develop new techniques.

References

Bibby, C. J., N. D. Burgess, and D. A. Hill. 1992. *Bird Census Techniques*. London: Academic Press.

Block, W. M., K. A. With, and M. L. Morrison. 1987. On measuring bird habitat: Influence of observer variability and sample size. *Condor* 72:182–189.

Bookhout, T. A. (ed). 1994. *Research and Management Techniques for Wildlife and Habitats*. 5th ed. Bethesda, Md.: Wildlife Society.

Conroy, M. J. 1996. Abundance indices. Pages 179–192 in D. E. Wilson, F. R. Cole, J. D. Nichols, R. Rudran, and M. S. Foster (eds.), *Measuring and Mon-*

itoring Biological Diversity: Standard Methods for Mammals. Washington, D.C.: Smithsonian.

Cook, C. W., and J. Stubbendieck (eds.). 1986. *Range Research: Basic Problems and Techniques.* Denver: Society for Range Management.

Cooper, C. F. 1957. The variable plot method for estimating shrub density. *Journal of Range Management* 10:111–115.

Corn, P. S. 1994. Terrestrial amphibian communities in the Oregon Coast Range. Pages 304–317 in L. F. Ruggiero et al. (tech. coords.), *Wildlife and Vegetation of Unmanaged Douglas-fir Forests.* General Technical Report PNW-GTR-285. Washington, D.C.: USDA Forest Service.

Crump, M. L., and N. J. Scott, Jr. 1994. Visual encounter surveys. Pages 84–92 in W. R. Heyer, M. A. Donnelly, R. W. McDiarmid, L. C. Hayek, and M. S. Foster (eds.), *Measuring and Monitoring Biological Diversity: Standard Methods for Amphibians.* Washington, D.C.: Smithsonian.

Dodd, M. G., and T. M. Murphy. 1995. Accuracy and precision of techniques for counting great blue heron nests. *Journal of Wildlife Management* 59:667–673.

Fellers, G. M., and C. A. Drost. 1994. Sampling with artificial cover. Pages 146–150 in W. R. Heyer, M. A. Donnelly, R. W. McDiarmid, L. C. Hayek, and M. S. Foster (eds.), *Measuring and Monitoring Biological Diversity: Standard Methods for Amphibians.* Washington, D.C.: Smithsonian.

Ford, H. A., L. Bridges, and S. Noske. 1990. Interobserver differences in recording foraging behavior of fuscous honeyeaters. *Studies in Avian Biology* 13:199–201.

Ganey, J. L., and W. M. Block. 1994. A comparison of two techniques for measuring canopy closure. *Western Journal of Applied Forestry* 9:21–23.

Gotfryd, A., and R.I.C. Hansell. 1985. The impact of observer bias on multivariate analyses of vegetation structure. *Oikos* 45:223–234.

Hatton, T. J., N. E. West, and P. S. Johnson. 1986. Relationships of error associated with ocular estimation and actual cover. *Journal of Range Management* 39:91–92.

Heyer, W. R., M. A. Donnelly, R. W. McDiarmid, L. C. Hayek, and M. S. Foster (eds.). 1994. *Measuring and Monitoring Biological Diversity: Standard Methods for Amphibians.* Washington, D.C.: Smithsonian.

Immelmann, K., and C. Beer. 1989. *A Dictionary of Ethology.* Cambridge, Mass.: Harvard University Press.

Jaeger, R. G. 1994a. Patch sampling. Pages 107–109 in W. R. Heyer, M. A. Donnelly, R. W. McDiarmid, L. C. Hayek, and M. S. Foster (eds.), *Measur-*

ing and Monitoring Biological Diversity: Standard Methods for Amphibians. Washington, D.C.: Smithsonian.

———. 1994b. Transect sampling. Pages 103–107 in W. R. Heyer, M. A. Donnelly, R. W. McDiarmid, L. C. Hayek, and M. S. Foster (eds.), *Measuring and Monitoring Biological Diversity: Standard Methods for Amphibians.* Washington, D.C.: Smithsonian.

Jaeger, R. G., and R. F. Inger. 1994. Quadrat sampling. Pages 97–102 in W. R. Heyer, M. A. Donnelly, R. W. McDiarmid, L. C. Hayek, and M. S. Foster (eds.), *Measuring and Monitoring Biological Diversity: Standard Methods for Amphibians.* Washington, D.C.: Smithsonian.

Jones, C., W. J. McShea, M. J. Conroy, and T. H. Kunz. 1996. Capturing mammals. Pages 115–155 in D. E. Wilson, F. R. Cole, J. D. Nichols, R. Rudran, and M. S. Foster (eds.), *Measuring and Monitoring Biological Diversity: Standard Methods for Mammals.* Washington, D.C.: Smithsonian.

Kepler, C. B., and J. M. Scott. 1981. Reducing count variability by training observers. *Studies in Avian Biology* 6:366–371.

Kuenzi, A. J., G. T. Downard, and M. L. Morrison. 1999. Bat distribution and hibernacula use in west-central Nevada. *Great Basin Naturalist* 59:213–220.

Kunz, T. H. (ed.). 1988. *Ecological and Behavioral Methods for the Study of Bats.* Washington, D.C.: Smithsonian.

Kunz, T. H., D. W. Thomas, G. C. Richards, C. R. Tidemann, E. D. Pierson, and P. A. Racey. 1996. Observational techniques for bats. Pages 105–114 in D. E. Wilson, F. R. Cole, J. D. Nichols, R. Rudran, and M. S. Foster (eds.), *Measuring and Monitoring Biological Diversity: Standard Methods for Mammals.* Washington, D.C.: Smithsonian.

Lancia, R. A., and J. W. Bishir. 1996. Removal methods. Pages 210–217 in D. E. Wilson, F. R. Cole, J. D. Nichols, R. Rudran, and M. S. Foster (eds.), *Measuring and Monitoring Biological Diversity: Standard Methods for Mammals.* Washington, D.C.: Smithsonian.

Lindsey, A. A., J. D. Barton, and S. R. Miles. 1958. Field efficiencies of forest sampling methods. *Ecology* 39:434–444.

Ludwig, J. A., and J. F. Reynolds. 1988. *Statistical Ecology: A Primer on Methods and Computing.* New York: Wiley.

Martin, P., and P. Bateson. 1993. *Measuring Behavior.* 2nd ed. Cambridge: Cambridge University Press.

Morrison, M. L., and L. S. Hall. 1999. Habitat relationships of amphibians and reptiles in the Inyo-White mountains, California and Nevada. Pages 233–237 in S. B. Monsen and R. Stevens (eds.), *Proceedings: Ecology and Management* of *Pinyon-Juniper Communities within the Interior West.* Proceedings RMRS-P-9. Ogden, Utah: USDA Forest Service, Rocky Mountain Research Station.

Morrison, M. L., B. G. Marcot, and R. W. Mannan. 1998. *Wildlife-Habitat Relationships: Concepts and Applications.* 2nd ed. Madison: University of Wisconsin Press.

Ralph, C. J., and J. M. Scott. 1981. Estimating numbers of terrestrial birds. *Studies in Avian Biology* 6:1–630.

Ralph, C. J., G. R. Geupel, P. Pyle, T. E. Martin, and D. F. DeSante. 1993. *Handbook for Field Methods for Monitoring Landbirds.* General Technical Report PSW-144. Washington, D.C.: USDA Forest Service.

Ralph, C. J., J. R. Sauer, and S. Droege. 1995. *Monitoring Bird Populations by Point Counts.* General Technical Report PSW-GTR-149. Berkeley: USDA Forest Service, Pacific Southwest Research Station.

Remsen, J. V., and S. K. Robinson. 1990. A classification scheme for foraging behavior of birds in terrestrial habitats. *Studies in Avian Biology* 13:144–160.

Reynolds, R. P., R. I. Crombie, R. W. McDiarmid, and T. L. Yates. 1996. Voucher specimens. Pages 63–69 in D. E. Wilson, F. R. Cole, J. D. Nichols, R. Rudran, and M. S. Foster (eds.), *Measuring and Monitoring Biological Diversity: Standard Methods for Mammals.* Washington, D.C.: Smithsonian.

Reynolds, R. T., J. M. Scott, and R. A. Nussbaum. 1980. A variable circular-plot method for estimating bird numbers. *Condor* 82:309–313.

Roberts, G., and P. R. Evans. 1993. Responses of foraging sanderlings to human approaches. *Behaviour* 126:29–43.

Roberts, W. A., and S. Mitchell. 1994. Can a pigeon simultaneously process temporal and numerical information? *Journal of Experimental Psychology: Animal Behavior Processes* 20:66–78.

Rosenthal, R. 1976. *Experimenter Effects in Behavioral Research.* New York: Irvington.

Rudran, R., T. H. Kunz, C. Southwell, P. Jarman, and A. P. Smith. 1996. Observational techniques for nonvolant mammals. Pages 81–104 in D. E. Wilson, F. R. Cole, J. D. Nichols, R. Rudran, and M. S. Foster (eds.), *Measuring and Monitoring Biological Diversity: Standard Methods for Mammals.* Washington, D.C.: Smithsonian.

Schleidt, W. M., G. Yakalis, M. Donnelly, and J. McGarry. 1984. A proposal for a standard ethogram, exemplified by an ethogram of the bluebreasted quail (*Coturnix chinensis*). *Zeitschrift für Tierpsychologie* 64:193–220.

Schultz, A. M., R. P. Gibbens, and L. DeBano. 1961. Artificial populations for teaching and testing range techniques. *Journal of Range Management* 14:236–242.

Scott, J. M., F. L. Ramsey, and C. P. Kepler. 1981. Distance estimation as a variable in estimating bird numbers from vocalizations. *Studies in Avian Biology* 6:334–340.

Scott, N. J., Jr. 1994. Complete species inventories. Pages 78–84 in W. R. Heyer, M. A. Donnelly, R. W. McDiarmid, L. C. Hayek, and M. S. Foster (eds.), *Measuring and Monitoring Biological Diversity: Standard Methods for Amphibians.* Washington, D.C.: Smithsonian.

Smallwood, K. S., and C. Schonewald. 1996. Scaling population density and spatial patterns for terrestrial, mammalian carnivores. *Oecologia* 105:329–335.

Szewczak, J. M., S. M. Szewczak, M. L. Morrison, and L. S. Hall. 1998. Bats of the White and Inyo mountains of California-Nevada. *Great Basin Naturalist* 58:66–75

White, G. C., and R. A. Garrott. 1990. *Analysis of Wildlife Radio-Tracking Data.* San Diego: Academic Press.

Wilson, D. E., F. R. Cole, J. D. Nichols, R. Rudran, and M. S. Foster (eds). 1996. *Measuring and Monitoring Biological Diversity: Standard Methods for Mammals.* Washington, D.C.: Smithsonian.

Zimmerman, B. L. 1994. Audio strip transects. Pages 92–97 in W. R. Heyer, M. A. Donnelly, R. W. McDiarmid, L. C. Hayek, and M. S. Foster (eds.), *Measuring and Monitoring Biological Diversity: Standard Methods for Amphibians.* Washington, D.C.: Smithsonian.

CHAPTER 7

Designing a Reserve

The design of conservation reserves has been widely debated for decades. The central issues in reserve design—genetics, demography, fragmentation, and corridors—have been amply discussed elsewhere. (See, for example, Margules and Usher 1981 and Margules et al. 1988.) This chapter explains how you can apply the principles of reserve design to the restoration of wildlife habitat—especially the major aspects of broad-scale (or landscape) issues. Topics range from selecting the restoration site to the broad design of the project.

Selecting a Site

The literature on selection algorithms is large and well defined and has remained relatively stable since the early 1980s (Margules and Usher 1981; Pendergast et al. 1999). The adaptive management literature, though more than 20 years old (Holling 1978), has not been advanced by empirical trial to the point where it provides step-by-step solutions for action in the absence of information. Thus anyone developing a reserve must recognize how far their plans deviate from the assumptions of the reserve selection systems they use. Moreover, they need to recognize that a proposed system that is unsubstantiated by research is, in essence, a management experiment. As noted in

Chapter 5, establishing a rigorous monitoring plan will advance our understanding of restoration successes and failures.

Academics and planners have sought reproducible, if not quantitative, systems for evaluating biological resources for preservation. Two concepts have held prominence: maximizing the number of taxa protected and using the minimum amount of area (or cost) to achieve a conservation goal (Margules et al. 1988; Church et al. 1996). Most algorithms of reserve selection are iterative (best additive value at each step in a series of choices) or optimal (best grouping given a set of constraints).

Margules and Usher (1981) and Usher (1986) have listed the criteria most frequently used to rank the relative value of units in reserve selection:

- Species diversity (number)
- Presence of rare species
- Representativeness of species or habitat
- Total area
- Naturalness and level of disturbance
- Potential utility (products and use)
- Education value
- Level of threats
- Known history (both ecological and human)

To these original criteria others have added:

- Complementarity—the unique contribution of a new unit to an existing reserve system (Margules et al. 1988; Pressey et al. 1994)
- Representativeness (beyond just species)—including ranges of physiographic variation (Austin and Margules 1986)
- Irreplaceability—the extent to which conservation options are altered if a site is lost (Pressey et al. 1994)
- Integrity—the capability of supporting "natural systems" (Angermeier and Karr 1994)
- Opportunity cost—the relative cost to management if a unit is lost (Pressey and Tully 1994)
- Threat—the likelihood that a unit will be destroyed (Rossi and Kuitumen 1996)
- Survivability—threat plus some estimate of resilience (Lockwood et al. 1997)

Underlying these criteria is the premise of reserve selection: persistence depends on the capacity of a reserve to correct for predictable, unpredictable, and unknowable problems.

Selection Criteria for Protecting Endangered Species

Methods for bundling groups of endangered species plans are poorly represented in the literature. In fact, criteria to judge the relative contribution of units to the maintenance of individual species are less well developed than their ecosystem counterparts, and the bulk of the literature covers habitat relationships rather than habitat planning (Morrison et al. 1998). In many cases the issues have been defined—such as source and sink populations, genetics (Schonwall-Cox et al. 1983:414–445), and population vulnerability and viability (Soulé and Simberloff 1986; Soulé 1991)—but the systems to turn these models into reserve selection criteria are less well developed (Caughley and Gunn 1996:217).

Selection Criteria for Protecting Ecosystems

The exercise of selecting habitat units or ecosystems to protect and restore comes closest to the overall goals of reserve selection algorithms. Caughley and Gunn (1996) suggest that this exercise can be focused on preserving states (species assemblages) or preserving processes (species and system interactions), each with its own set of objectives. Neither state-focused nor process-oriented studies, however, escape the biases or management problems caused by incomplete information about the study area. With sufficient information, process preservation deals with persistence during selection of reserves whereas state preservation tends to shunt the issue of persistence onto the subsequent management (such as restoration). Defining and predicting ecosystem processes, specifically disturbance events, remain an issue (Orians 1993). The newer criteria—integrity, opportunity cost, threat, survivability—appear to be a better match for process preservation than the older criteria, particularly in disturbed landscapes. The notion that post facto restoration or other management practices can solve the problem of reserve designs remains untested. Reserves that require such post facto corrections are more difficult to maintain than those with functional ecosystems.

Special Considerations

One of the most important features of a selection algorithm is its ability to describe the complementary attributes of flexibility and irreplaceability. Pressey et al. (1994) define flexibility as the capacity of a study area to produce a number of alternative reserve configurations that meet specific conservation strategies; they estimate the irreplaceability of a planning unit by the number of times it appears as a component part of different alternatives.

Estimates of flexibility and irreplaceability in a study area can focus discussions about the importance of a specific unit or reserve design alternative. When reserve system goals are unclear, estimates of irreplaceability offer a feedback mechanism for setting levels of resource protection. A geographic information system (GIS) becomes invaluable for quick modeling of reserve alternatives and subsequent identification of irreplaceable areas.

The distribution of biological resources determines the success and efficiency of a selection system. Lack of uniformity in the distribution of sensitive species and habitats—as well as lack of uniformity in the information available—confound most selection processes. One of the strongest attributes is the degree to which the distribution of rare species or sensitive habitats is nested within sets of common species. Ryti and Gilpin (1987) point out that rare plants in southern California are not nested within the occurrence of all plant species; therefore a reserve system based on the maximum number of plant species will be relatively inefficient in protecting them. The occurrence of narrow endemics in areas of low species richness appears to be a general pattern in most Mediterranean ecosystems and is perhaps the case for most North American endemics. The degree to which rare species are not nested within other species distributions strongly affects reserve selection processes and was one of the contributing factors in the debate over one large reserve versus several small ones (the "SLOSS" debate).

Most planning processes fail to collect information on all the species or ecosystems under consideration. The short time frame of planning (1 or 2 years) leads many biologists to use the information at hand rather than collect new data. The danger is not just the absence of information; rather, it is the interpretations that go beyond what is defensible and the mixture of questionable data with reliable data. Indicator species have been proposed as a means of overcoming the limited data on sensitive species in project planning. Indicators of any kind must be carefully and narrowly defined if they are to have any predictive value (Landres et al. 1988; Morrison et al. 1992). Flather et al. (1997) have shown that one taxon usually fails to predict the other groups. Planners still use—some would say misuse—indicator taxa because they believe there are no alternative methods (Roberts 1988). Mountain lions (*Felis concolor*) and golden eagles (*Aquila chrysaetos*) have been used in southern California because both are sensitive to habitat fragmentation. Yet both can tolerate a degree of vegetation alteration and continue to use areas that have become unacceptable to the majority of rare species.

Wildlife/habitat relationship (WHR) models have been used to estimate wildlife occurrence in the absence of information. Morrison et al. (1998) have reviewed this topic in detail; two of their most consistent suggestions are to identify your objectives clearly and validate your model each step of the way. In many cases, species have been linked to geographic units (typically vegetation polygons in a GIS) as a substitute for empirical data on distributions. WHR models at the geographic scale are difficult to test; species distributions have historical artifacts and temporal variation that are transparent to static maps of vegetation; and the hypothesis that populations respond in the same manner to the same set of variables across a species' range is tenuous at best. On a landscape scale, the translation of point observations of species occurrence to polygons of distribution is functional at the statewide scale of gap analysis (Scott et al. 1993) but requires a great deal of interpretation and discretion at the localized scale of many restoration projects.

The principles of reserve design can be summarized as follows from Noss et al. (1997):

- Species well distributed are less susceptible to extinction than species confined to small locations.
- Large blocks, containing large populations, are better than small blocks.
- Blocks of habitat close together are better than blocks far apart.
- Habitat in continuous blocks is better than fragmented habitat.
- Interconnected blocks of habitat are better than isolated blocks.
- Populations that fluctuate are more vulnerable than stable populations.
- Disjunct or peripheral populations are likely to be more genetically impoverished and vulnerable to extinction—but also more genetically distinct—than central (core) populations.

As noted in Chapters 1 and 2, animal movements (dispersal, migration) and specific resource requirements and constraints are fundamental aspects of restoration. And as we have seen, the success of a restoration project depends on the spatial context—the landscape perspective—of the animals in question and how the available land meets their requirements (Bissonette 1997).

Size

A planning unit of insufficient size will not be a viable reserve for the species being assessed. But how large is large enough? Large areas are better than small ones for maintenance of species. The basic species/area relationship simply means that the larger the area, the more species it can accommodate. Moreover, larger areas are capable of maintaining species for a longer period

FIGURE 7.1.
Isolated, small reserves, such as the one depicted here in San Diego County, California, seldom enhance the restoration of wildlife. (Photo courtesy of Thomas A. Scott.)

of time. Studies from across North America have shown that isolated patches smaller than 100 ha usually do not retain their complement of native vertebrate species for longer than a few decades (Figure 7.1). Soulé et al. (1992) found evidence that large predators (coyotes, foxes) retard the biotic collapse of these small remnants by controlling populations of smaller predators such as domestic and feral cats. Plant species may disappear, too, because of chronic and cumulative disturbances and changes in fire frequency.

The size of the area needed to maintain native fauna may be based on habitat quality. An area with low-quality resources for a species of interest may need to be larger than an area with higher-quality resources (type and amount of food, for example, number of predators) (Meffe and Carroll 1997:313–314). A planning unit's size has implications for the amount of edge as well and hence the expected degree of edge effect. Small areas have a greater edge/interior ratio, thus increasing edge and reducing the amount of interior. This shift may make the area more vulnerable to invasion by undesired plants and animals, and it increases the potential harm of environmental influences such as temperature and wind (Meffe and Carroll 1997:316). Typical examples in restoration planning would be the narrow border of vegetation surrounding a pond or the greenbelt surrounding a housing development. Huxel and Hastings (1999) have reviewed the importance of a population's spatial dynamics in restoration planning. They note that either restoring patches adjacent to occupied patches or reintroducing the species into restored patches increases the efficacy of the recovery effort.

Heterogeneity and Dynamics

The internal structure of an area, including species composition and population densities and dispersions, is defined by the pattern of disturbance and longevity of the patches. Regardless of its size, the planning area is thus a mosaic of patches of various sizes and ages and its faunal diversity depends on the number of these patches and their dynamics. Remember that a "patch of habitat" is a species-specific concept and must therefore be viewed in light of the requirements of individual species (the patch size of a salamander, for example, versus that of a rabbit). Not only do patches change through time, but new patches are created by succession and disturbance and thus their spatial relationships change.

Therefore, you must consider the desired heterogeneity—as well as the desired dynamics of each patch—of a project area during restoration planning. Ideally the restored area will be large enough to include the natural patches required by the target species. Recall that many populations may form metapopulations (Chapter 1). The size of most restoration projects is predetermined, however, and will be too small to contain metapopulation structure for any but the smallest vertebrates. In such cases you should consider the patch relationships of specific animal species. In small reserves, the disturbance needed to maintain the vegetative component of a species habitat in a certain seral stage could result in its extinction (because no undisturbed refugia were available). In many cases you will have to substitute an alternative treatment for the natural disturbance—using hand grubbing, for example, in lieu of fire.

Small vertebrates may, in fact, be more susceptible to changes in metapopulation structure than the larger species (megafauna) that have been publicized as umbrella species under which the smaller species are protected. Small species, with their short generation times, high rates of population increase, and high habitat specificity, are more vulnerable to localized density-independent environmental factors than are the megavertebrates (because they are more dependent on smaller areas and usually have less mobility).

Landscape Context

The biotic components of a reserve are influenced by the patch dynamics within the reserve as well as by influences from the surrounding area—that is, the landscape matrix. The reserve can be seen as a patch within a larger spatial extent. As Meffe and Carroll (1997) explain, some species use different patches seasonally and must be able to locate these patches within the larger landscape matrix. Migratory herbivores (deer, elk), for example, move to low-elevation areas in winter and to higher mountain meadows during the summer. Thus they move between reserves. In contrast, some salamanders develop as larvae in ponds and move to adjacent upland areas as adults, thus confining their movements within a single reserve.

George and Zack (2001) have reviewed the influence of large-scale processes on the distribution and abundance of wildlife in a restoration context. For example, they discuss the regional effects of road density on the habitat use of wolf (*Canis lupus*) packs (Mladenoff et al. 1999). Most areas with road densities greater than 0.45 km/km² did not have wolves. Some areas predicted to provide wolf habitat were not occupied, however, because the wolves had not had sufficient time to colonize the entire region. This demonstrates that often you must take a regional approach to habitat for successful restoration of animals with large use areas.

Even animals with small use areas may be influenced by regional landscape patterns. Predation of songbird nests, for example, appears to be influenced more by forest cover at large scales (10,000 km²) than at smaller spatial scales. Thus your restoration plans for small songbirds should consider regional landscape patterns. For mobile species, the temporal duration of habitat may be more important than the distance between patches of habitat. Drought reduces food availability over the short term for many species, for example, but periodic droughts prevent successional changes in vegetation that would eventually make an area unsuitable for the species. The restoration strategy is to create links between habitat patches and ensure that all patches are not unsuitable at the same time (George and Zack 2001). The habitat surrounding a restoration site may also influence the success of restoration. Birds nesting in habitat surrounded by human-altered areas, for example, experience higher nest predation than those nesting in patches adjacent to natural areas (George and Zack 2001).

Pressey and Cowling (2001) list five stages of conservation planning:

• Identify your conservation goals for the planning region. This goal is necessarily subjective but sets the broad landscape context for specific restoration and other management activities. Targets are set for minimum area size, distances between restored areas, location and type of corridors, and so forth.

- Review the present conservation areas. Here you match what is needed (the preceding step) with what already exists.
- Select additional conservation areas. List the additional areas (by using various algorithms of reserve design, for example) that are necessary to meet your conservation goals, including those locations in need of restoration.
- Implement conservation actions. Here you apply the steps needed to meet the conservation goals (restoration, purchase, change in land-use practice).
- Maintain required values of conservation areas. This step involves the maintenance and monitoring of the conservation areas to ensure that your goals are met.

If the regional landscape mosaic changes—through fragmentation or changes in land-use practices, for example—the dynamics between patches of habitat will change. Certain patches will be highly productive (*sources*) and produce surplus animals that may settle in less productive patches (*sinks*). As depicted in Figure 7.2 (Wiens 1989:fig. 4.12), populations in

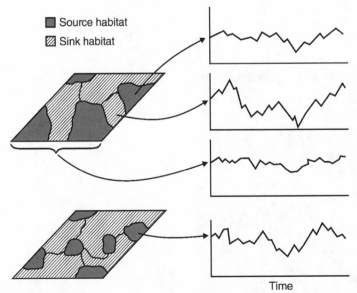

FIGURE 7.2.
Representation of the dynamics of populations (or a metapopulation) in patches of a landscape mosaic. (From J. A. Wiens, *The Ecology of Bird Communities, Vol. 2: Processes and Variations,* Figure 4.12. Page 174. Copyright 1989. Reprinted with permission of Cambridge University Press.)

sources may vary less than those in sinks, which might become extinct at a local scale. At a regional scale summed over patches that are following independent dynamics, the population (or metapopulation) might be more stable. For the restorationist, determining the dynamics of a population on a landscape scale reveals priority locations for enhancement (changing a sink to a source, for example).

Corridors

It is generally thought that population viability is enhanced when individuals of a subpopulation are able to move to another subpopulation. (Recall the discussion of metapopulations in Chapter 1.) At some spatial scale, all locations are heterogeneous (patchy). What may appear to be homogeneous to a large mammal may be quite heterogeneous at the spatial scale of the amphibian. Yet there is always a larger spatial scale at which the landscape becomes patchy even for a wide-ranging large mammal. Often it is the connectivity among patches that determines whether a species can survive in a specific location—and perhaps even within a larger region (Figure 7.3).

As Beier and Noss (1998) point out, until recently most species lived in well-connected landscapes. But human impacts such as urbanization, agriculture, and road construction have disrupted connectivity. Thus many con-

FIGURE 7.3.
Riparian, estuarine, and coastal chaparral environments combine to form the wildlife corridors that extend from the sea into the foothills and mountains of the Coastal Range of southern California. (Photo courtesy Zoological Society of San Diego.)

servation biologists have been advocating the preservation or development of corridors to link landscape patches. Moreover, maintaining pathways for dispersal among subpopulations is an important consideration in managing the genetic composition of a regional population (Mech and Hallett 2001).

The intuitive appeal of corridors has resulted in a widespread recommendation for their use in conservation planning and landscape design. A classic example of a reserve designed to promote population viability is shown in Figure 7.4. Note the corridor extending out of the central part of the reserve (the core). The concept behind this basic corridor design is to allow movement of animal species of all sizes. First, the corridor represents a passage route for species that can live in the core (or patch) but are too large to have their home range contained within the narrow corridor—as for as

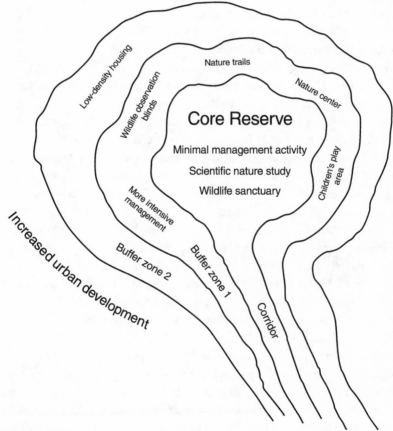

FIGURE 7.4.
Typical design for an urban wildlife reserve.

medium-sized and large mammals, for example. Second, the corridor represents an extension of the core that allows some degree of residency if not breeding opportunities—for small mammals and small to medium-sized birds, for example.

The use of corridors, however, has been met with much controversy. As Hess (1994) observes, corridors can sometimes increase the chance of metapopulation extinction by promoting the transmission of disease. Thus you should consider the probability of disease and other adverse factors (such as parasites) during the planning stages of all reserve networks. Strategies for containing disease should be developed *before* the links are established. Contingencies for treating epizootics include vaccination, removal of infected individuals, and temporary termination of the link.

Simberloff and Cox (1987) and Simberloff et al. (1992) have questioned the value of corridors. We do not have information, they say, available to warrant wholesale adoption of corridors as a conservation action. In addition to disease, corridors may enhance the spread of fires and other catastrophes, and expose individuals to predation, domestic animals, and poachers. And corridors obviously take financial resources that might be better spent on another aspect of reserve establishment. Thus Simberloff and his colleagues recommend that each proposed corridor be evaluated on its own merits because species/environmental interactions differ in time and space.

Simberloff et al. (1992) have outlined several alternatives to corridors that could maintain species in a project area. Managing the landscape as a whole, they suggest, rather than managing each reserve as an independent unit, could reduce the need for corridors. (Maintaining patches of vegetation surrounding a core area in different seral stages, for example, could allow movement without the need for a linear corridor.) Another alternative is the development of "stepping stones" that are close enough to allow animals to move between patches without the need for direct links.

Beier and Noss (1998) have defined corridors with regard to "specific wildlife populations" (see also Simberloff and Cox 1987). They note that evaluating the utility of corridors "only makes sense in terms of a particular focal species and landscape." This conclusion follows naturally from the notion of habitat as a species-specific phenomenon. In the species-specific context, then, we can define a corridor as a linear habitat, embedded in a matrix of dissimilar habitats, that connects two or more larger blocks of habitat. Passage is defined as travel via a corridor from one habitat patch to another. Beier and Noss explicitly exclude linear habitats that support many species but do not connect large habitat patches (such as narrow riparian strips embedded in agricultural fields). Although I understand their rationale, they seem to have overlooked the spatial nature of patchiness in making this distinction. As we have seen, individuals with small home ranges but long dispersal distances may perceive patchiness at the forest floor level—where changes in soil moisture or ground cover could substantially change the fine-scale patchiness of such an animal's landscape. Thus planning for corridors requires designs at multiple spatial scales depending on the overall goals of the study and species involved. As Diefenbach et al. (1993) point out, restorationists need to study all aspects of the environment that animals might encounter.

Empirical Evidence

Because corridors have the theoretical potential to enhance population viability, the fundamental question you might ask is: "Are animals passing through the corridor?" Once passage is confirmed, however, you must determine whether the animals are using the corridor sufficiently often to influence population viability. This means finding out the movement rates required to maintain genetic diversity and prevent local extinctions among the habitat patches. Not only is gathering such information beyond the means of most projects, but some studies are indicating that it is nearly impossible to develop adequate models of populations that occupy disjunct habitat patches (that is, spatially explicit models). Nevertheless, the literature offers estimates of such parameters for a wide variety of wildlife species and, further, surrogates for the ideal parameters may sometimes be appropriate. Sex and age ratios, territory occupancy, and recruitment rates, for example, may be adequate surrogates for direct measures of population survival and fecundity for this application (Morrison et al. 1998).

You will have to determine, as well, whether animals are using a presumed corridor at a greater frequency than they are using the surrounding landscape matrix. That is: measuring movement through a human-defined corridor without examining other movement paths could lead to the false conclusion that the corridor is essential to the species. Beier and Noss (1998), for example, found that many studies fail to address movement through the matrix. After reviewing the literature between 1980 and 1997 to assess the state of knowledge regarding the value of corridors as passage routes, they categorized each study by the types of parameters it measured (population parameters, movements of individual animals, or the putative hazards of corridors) and whether it was observational or experimental. Because most of these studies were found to contain "design limitations," they found only 12 studies that could be used to address their objective. Beier and Noss concluded that 10 of these studies offered persuasive evidence that corridors provided sufficient connectivity to improve the viability of populations. They also found that many of the purely observational studies offered evidence of animal movements through corridors.

Despite the importance of quantifying animal movements within and between corridors in reserve design, little research has been conducted on animal dispersal. As Van Vuren (1998) notes, behavioral ecologists have been studying dispersal only since the mid-1970s. And although much of this work has focused on the role of dispersal in population processes—such as population regulation and population genetics—dispersal is fundamentally an attribute of individual animals and has important implications for individual fitness. To concentrate on dispersal as a population-level phenomenon, therefore, obscures the individual nature of dispersal.

Usually defined as the one-way movement of an animal away from its current home range, dispersal typically involves juveniles or young adults. In most vertebrates, dispersal is skewed toward males. Sex-based dispersal has important implications for reserve design and restoration because dispersers that reach a new area may be predominately of one sex (Van Vuren 1998). Although the factors determining where a dispersing animal settles are complex, the common theme, both theoretically and empirically, indicates that an individual keeps moving until it locates habitat where it can settle. Recall that habitat entails far more than vegetation and that the proper resources must be present. Other factors, such as availability of females, influence settlement. If such factors are not present, dispersers will not settle in the area desired by the restorationist.

The direction of dispersal may be influenced by topographic features. For wide-ranging species, canyons, lakes, and rivers will influence the direction of movement. Animals ranging less widely, perhaps just within the reserve, may be influenced by such factors as soil moisture and ground cover. Dispersal direction, as well as barriers to dispersal, must be evaluated for each species. In fact, some species are known to disperse across what seem to be unsuitable areas. Shrews (*Sorex*) and mice (*Peromyscus* spp.), for example, are known to travel more than 600 m across frozen lakes. When Van Vuren (1998:table 14-1) summarized dispersal distances in mammals, he found a close positive correlation between body mass and dispersal distance. Median distances ranged from less than 100 m to 1.5 km in rodents, lagomorphs, and small carnivores; from 2 to 10 km in many medium-sized carnivores; and from 10 to 65 km in the larger mammals.

Although we have little information on the survival of dispersing animals, the evidence indicates that their increased movements make them more vulnerable to predation. Moving through unfamiliar terrain is also likely to increase mortality. As Van Vuren (1998) concludes, even a reserve system designed to be well within a species' dispersal capabilities will fail if most of the dispersers die while attempting to move.

Case Studies

Haas (1995) found that the movements of three migratory bird species among riparian woodlands and shelterbelt woodlots in North Dakota, although rare, occurred more frequently between sites connected by wooded corridors than between unconnected sites. He concluded that knowledge of patterns of fledgling, natal, and breeding dispersal of birds in patchy environments would substantially aid our decisions on reserve design and habitat corridors.

As we have seen, there are troubling aspects of corridor use for promoting population viability (such as disease transmission and population traps). In Australia, Downes et al. (1997) found that exotic black rats (*Rattus rattus*) were abundant in corridors and that this abundance might be jeopardizing movements of the native bush rat (*R. fuscipes*). Although several other studies have addressed the negative impacts of corridors (Bennett 1990; Seabrook and Dettmann 1996; Stoner 1996), none seems to provide strong empirical evidence of such impacts (Beier and Noss 1998). It seems reasonable to assume, however, that exotic predators (domestic and feral cats, dogs, rats) could have a severe impact on certain species attempting to use a corridor in an urban environment.

Certain types of narrow corridors may attract nest parasites such as the brown-headed cowbird. As Robinson et al. (1995) have shown, numbers of cowbirds and rates of parasitism may be higher within corridors because they are essentially all edge. Further, certain corridors, especially riparian areas, may have high densities of host species, which further attracts cowbirds. It could be argued (Beier and Noss 1998) that many of these situations are not true passage corridors but isolated, narrow habitats. Nevertheless, birds that are drawn to nest in such areas may experience poor reproductive success. This is a complicated issue. After all, you might ask, where would the birds have nested without these corridors? As Simberloff et al. (1992) point out, it might be better to focus our conservation efforts on establishing larger habitat areas rather than worrying about linking small sites.

Biologists think that riparian buffer strips may attract predators, perhaps at undesirable density and diversity. Vander Haegen and DeGraaf (1996) found higher predation rates on bird nests, both on the ground and in shrubs, in riparian buffer strips created by commercial clear-cutting than in intact forests. Predation rates were similar in mainstem and tributary buffer strips. The predators consisted of six mostly forest-dwelling species that used the buffers to forage and perhaps travel. The authors recommend buffer strips at least 150 m wide along riparian zones to reduce edge-related nest predation—especially in landscapes where buffers comprise a significant portion of the remnant forest.

Corridors may also include habitat components more or less linearly arranged across a landscape, such as perch poles in the desert. In the Mojave Desert of California, Knight and Kawashima (1993) found higher densities of common ravens (*Corvus corax*) along highways and powerlines than in control areas (no highways or powerlines within 3.2 km), and raven nests were more abundant along powerlines. Ravens may have been attracted to highways for road-kill carrion. Red-tailed hawks (*Buteo jamaicensis*) and their nests were more abundant along powerlines than along highways or control transects. The authors recommend that land managers evaluate the possible effects on vertebrate populations and species interactions when assessing future linear right-of-way projects.

Another form of corridor is transmission-line cuts, which may open forest or woodland canopies, provide lush grass, forb, or shrub cover, and create linear edge across a landscape that intersects other resource patches. Chasko and Gates (1982), for example, found that in a Maryland oak-hickory forest the corridor was dominated by mixed-habitat bird species—species that use two or more vegetation conditions such as grassland and shrubs—rather than grassland birds. They also found that the few isolated shrub patches occurring in the grassy corridor provided "habitat islands" where nest density and fledging success were high. Apparently predators were unable to exploit patchily distributed shrub nests in the corridor. Therefore, the authors recommend managing for increased vegetation heterogeneity within transmission-line corridors to increase nest density and success of mixed-habitat bird species.

Corridors and habitat connections have been proposed for large mammals as well. Silvicultural prescriptions designed to provide deer habitat corridors have been popular for some time (Wallmo 1969). Beier's (1993) simulation study of cougar habitat merged consideration of minimum habitat area and corridor use. Beier concludes that habitat areas as small as 2200 km^2 could support cougars with low extinction risk if demographic rescue rates along habitat corridors were on the order of one to four immigrants per decade. Studying arboreal marsupials in southeastern Australia, Lindenmayer and Nix (1993) found that it is not enough to design wildlife habitat corridors based on suitable habitat, species home range, and predictions of minimum corridor width alone. They suggest the design criteria should also include site context, connectivity, and the social structure, diet, and foraging patterns of desired species.

Habitat isolates, corridors, and connectors do not, however, exist in static form. In most ecosystems, they are subject to systematic changes and random disturbances that may render their long-term conservation a major challenge. Morrison et al. (1998:chap. 9) discuss the dynamics of habitats in landscapes in more detail.

Research Needs

Beier and Noss (1998) cite two central areas of research that should be pursued to clarify the value of corridors. (Remember, however, that the value of corridors must be assessed by species.) First, they suggest that experiments using demographic parameters as dependent variables, even if unreplicated, should be done to demonstrate the demographic effect of particular corridors in particular landscapes. Second, observations of movements by naturally dispersing animals in already fragmented landscapes would demonstrate the conservation value of corridors if actual travel routes in both corridors and the surrounding matrix are analyzed. This point reiterates the importance of analyzing a restoration plan in the context of landscape and animal mobility (population structure). (See also Marzluff and Ewing 2001.)

Buffers

As Meffe and Carroll (1997) have noted, you can sometimes use land-use zoning to influence activities surrounding the reserve and make them more compatible with your project's goals. The classic diagram of a buffer applied to a reserve is depicted in Figure 7.5. Here activities in the buffer zone would be designed to minimize impacts on the core area. If logging is allowed in the transition zone, for example, then perhaps only camping should be permitted in the buffer. Or hunting might be allowed in the transition zone but prevented in the buffer and core. It can be argued, of course, that the core area should simply extend to the edge of the transition zone—thus providing at least the opportunity for a greater extent of the core. This is a valid argument but one that is often difficult to implement because of limitations

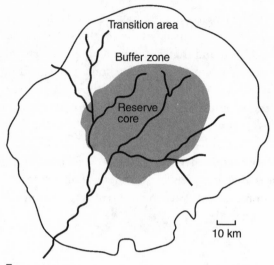

FIGURE 7.5.
Schematic diagram of a zoned reserve system showing the possible spatial relationship among a core area, a buffer zone, and a transition area. Human activities in each zone should match the goal of the zone and above all protect the reserve core. (From G. K. Meffe and C. R. Carroll, *Principles of Conservation Biology*, Figure 10.24. Copyright 1997, Sinauer Associates, Inc.)

in existing core area as well as of political considerations. The buffer can, in essence, be extended linearly from the reserve to connect with an adjacent reserve, thus forming a passage corridor.

Isolation

An important effect of isolation is the depression of species diversity. With small populations, isolation may increase the likelihood of adverse genetic effects such as fixation of deleterious alleles, increasing homozygosity, and overall decline in allelic diversity of the gene pool caused by genetic drift. Inbreeding depression—including depressed fertility and fecundity, increased natal mortality, and decreasing age of reproductive senescence—is one manifestation of small population size. Small colonizer populations are subject to founder effects that set the stage for loss of genetic and phenotypic diversity with subsequent isolation from outbreeding.

At the community level, isolation may result in a decline in species richness. When wildlife species are lost from isolated environments, this is termed *faunal relaxation*. This has been documented for oceanic islands isolated from mainlands by submergence of land bridges, for patches of forest isolated by slash-and-burn or clearcut timber harvesting, and for patches isolated by human development. The proportion of native plant cover in a chaparral fragment, for example, decreased over time in southern California (Figure 7.6). The edges of fragments are encroached by exotic plants, refuse, and other impacts—thus reducing the quantity and quality of vegetation and the habitat of many species within the fragment. Eventually the fauna reaches an equilibrium point where local extinction and emigration of species equal immigration and colonization. Although species may be added to such degraded fragments, they tend to be exotics (house mouse, European starling, numerous plants).

Isolation of reserves and parks has been a concern for biologists who suspect that relaxation effects are causing declines in native wildlife in protected areas. In some cases, legal and ecological boundaries of protected areas, such as national parks, do not necessarily align. It is imperative in such analyses to determine the actual status of the environment and populations in lands adjacent to the protected areas. Newmark (1986) has analyzed extinction of mammal populations in national parks in western North America and concludes that extinction rates exceed colonization rates (especially in small park units). He also found that two major factors cause greater rates of extinction

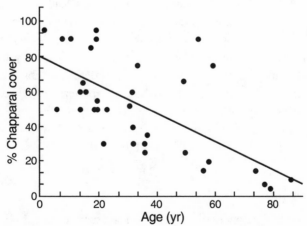

FIGURE 7.6.

Change in the proportion of native plant cover in isolated habitat remnants in San Diego County, California. (From M. E. Soulé et al., "The Effects of Habitat Fragmentation on Chaparral Plants and Vertebrates," Figure 3. *Oikos* 63:39–47. Copyright 1992.)

of mammals: small initial population size and short age of maturity (which is positively related to shorter population generation time).

Related to faunal relaxation is the delayed loss of competitive, rare species in landscape patches. The loss may take place over a period of time following isolation or alteration of environments as extinction eventually catches up with the conditions. Loehle and Li (1996) call this phenomenon *extinction debt* and think that it should be accounted for in conservation and reserve design. Isolation of environmental conditions in continental settings also contributes to development of *relictual faunas*—ancient species persisting because of suitable environments. By definition, relicts persist in refugia and may belong to any taxonomic group. Welsh (1990) has reported that the Del Norte salamander (*Plethodon elongatus*), Olympic torrent salamander (*Rhyocotriton olympicus*), and tailed frog (*Ascaphus truei*) are paleoecological relicts that have long been associated with ancient primeval conifer forests of the Pacific Northwest. Marcot et al. (1998) have compared the extant fauna of the interior western United States with Tertiary fossil fauna and report the persistence of relicts of 7 Tertiary genera (represented by 32 extant species) and 20 Tertiary families (represented by 55 extant genera). Unlike the relict salamanders reported by Welsh, the relict genera and families of the interior West occupy a wide range of environmental conditions including native grassland, shrubland, and forest.

It may be tempting to use relictual faunas as management or ecological indicators, but certain caveats must be addressed. Because relicts tend to occur in odd and disjunct locations, they do not necessarily represent zonal or climatic climax conditions. And insofar as relicts are holdovers from earlier environments, their distribution does not always reflect the suitability of current conditions. Thus you should be wary of defining habitat and landscape requirements based on current habitat-use patterns of relicts without knowing the paleoecological history of the population and the site. Nonetheless, plant and animal relicts are often of scientific interest and deserve special consideration (Millar and Libby 1991). See also Chapter 3.

Fragmentation

What we consider "fragmented" depends on the spatial scale of observation. It makes sense to view fragmentation in the context of how we view the selection of habitat by animals. Habitat selection is often viewed as a hierarchical process (see Chapter 2). First individuals select a broad geographic range, a decision that is largely innate. Within the geographic range the individual then makes a series of decisions based on increasingly refined combinations of vegetation structure, plant floristics, food resources, nest sites, and so on (Johnson 1980; Hutto 1985). Thus fragmentation is a crude concept unless the response of animals is placed in an appropriate spatial hierarchy. Changes take place in the environment at several scales of resolution (see also Angelstam 1996): fragmentation in a broad sense that takes place on the landscape scale; fragmentation of different plant associations that takes place within a vegetation type; and alteration of habitat quality within small areas.

There are numerous factors to consider when evaluating the effects of fragmentation on animals: (1) increased distances for movement paths; (2) loss of habitat area; (3) increased edge, less interior; (4) increased penetration of predators, competitors, and nest parasites; and (5) changes in microclimate. Item 1 is a macroscale issue; items 2, 3, and 4 operate at mesoscales; item 5 can operate at both meso- and microscales. When Bolger et al. (1997) studied the response of birds in a matrix of remaining chaparral vegetation in a residential landscape in southern California—a situation much more typical of birds studied in eastern locations—they found that 6 species showed negative (area-sensitive) responses, 4 species showed positive responses, and 10 species showed no apparent response to area size. In an Illinois grassland, 5 species were area-sensitive, 3 species responded positively to edge, and 6 species were restricted to specific vegetation configurations and were not area-sensitive (Herkert 1994). In their summary of North American studies, Freemark et al. (1995) concluded that the species present in a particular landscape setting showed negative, positive, and no responses to area size. They found that areas of under 10 ha are unsuitable for many species, whereas patches of 50 to 60 ha contain up to one-half of the regionally area-sensitive species. Often patch size had to reach 100 to 300 ha before area-sensitive species were present. But sensitivity to fragmentation and area size is reduced when more than 30 percent of the surrounding area is forested. Likewise in patches with interconnected vegetation, studies have found few species with a significant relation to patch size.

The factors influencing the response of wildlife to landscape fragmentation are numerous and interacting:

- Landscape context
- Forest cover
- Metapopulation structure and recolonization
- Forest type
- Patch size

- Patch shape
- Interpatch distance
- Edge length and composition
- Species' natural history
- Patch composition and habitat proportion
- Vegetation structure
- Predator abundance and composition
- Parasite abundance and composition
- Competitor abundance and composition
- Microclimate

How can you evaluate these factors when designing a wildlife-habitat restoration project? Most of these influences can be measured separately, but no single measure captures every aspect of shape and fragmentation. For example, patch size and shape, edge length, and forest type do not take isolation into consideration. These measures do not differentiate between degradation of an isolated patch and degradation of a patch of equal area and shape that has a close neighbor (Figure 7.7).

Marzluff and Ewing (2001) believe that restorationists can help maintain native birds in fragmented landscapes by performing a combination of short- and long-term actions designed to restore ecological function (not just shape and structure) to fragments:

- Maintaining native vegetation and deadwood in fragments
- Managing the landscape surrounding the fragment (the matrix), not just the fragment
- Making the matrix more like the native fragment
- Designing buffers that reduce penetration of undesirable species from the matrix
- Realizing that fragments might be best suited to conserve only a few species
- Developing monitoring programs that measure animal fitness

FIGURE 7.7.
Patches A, B, and C represent equal areas of habitat. Eliminating a patch leaves a landscape with equal patch area, interior area, and edge/area relationships. Fragmentation indices based on area or edge/area relationships do not reveal that a landscape of patches A and B is less isolated than a landscape of A and C or B and C. (From C. Davidson, "Issues in Measuring Landscape Fragmentation," Figure 1. *Wildlife Society Bulletin* 26:32–37. Copyright 1998.)

Different species are affected by different aspects of fragmentation. For mobile species with large home ranges, total habitat area is probably more important than interior area. The persistence of less mobile animals with small home ranges, by contrast, is likely related to patch isolation. Davidson (1998) has developed several guidelines for incorporating fragmentation issues into research and planning:

- The choice of spatial scale (both extent and grain) greatly affects the development of restoration plans. There is no such thing as the correct spatial scale for analysis; efforts to minimize fragmentation at one scale may increase it at another scale.
- Analyses are affected by the number of patch types and patch definitions. Evaluating fragmentation of an oak woodland, for example, will likely produce different conclusions than an analysis that broke the same woodland into blue oak and coast live oak woodlands.
- Perimeter/area ratios should not be used as measures of fragmentation. These ratios change unpredictably and do not capture important aspects of fragmentation such as isolation (Figure 7.8).
- It is better to measure different aspects of fragmentation separately than to try to select the "best" measure or combine them all into a single index.

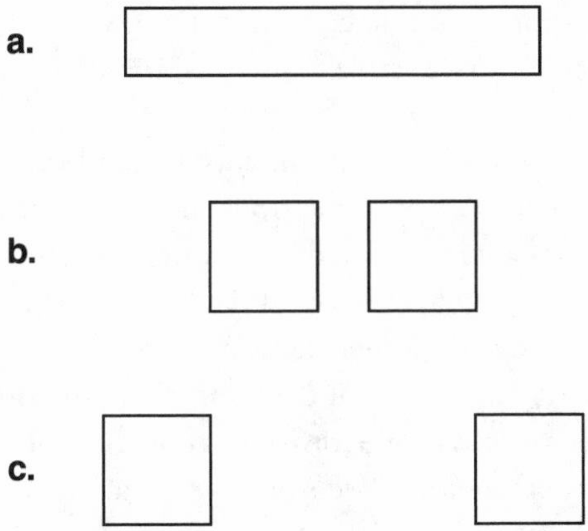

FIGURE 7.8.
The rectangle (part a) and the two squares (parts b and c) have the same total area and same total perimeter length. Assuming that the areas are large relative to a 120-m edge depth, the Mount Hood fragmentation index would assign the same value to all three landscapes. A perimeter/area ratio would also fail to distinguish between the three landscapes. Note: The areas in the figures are not drawn exactly to scale. (From C. Davidson, "Issues in Measuring Landscape Fragmentation," Figure 3. *Wildlife Society Bulletin* 26:32–37. Copyright 1998.)

Are Isolation and Fragmentation Always Bad?

Isolation of environments and populations is not inherently bad. In fact, there are circumstances where isolation is an advantage for conservation. One advantage is avoiding the spread of disease (Hess 1994), parasites, and pathogens. Another advantage is the establishing of several founder populations in sites with different disturbance dynamics and thus different likelihoods of success. In this case, overall persistence of the metapopulation is enhanced if there is little correlation of potentially disastrous environmental disturbances among the population centers, if populations are large enough to avoid genetic problems, or if there is occasional gene exchange (outbreeding) among populations. Yet another advantage of isolation is the maintaining of relictual faunas naturally isolated by changes in climate, vegetation, or landform.

In many cases, naturally developed faunas are best left isolated from exotic species—especially introduced predators and competitors. Species at the edge of their range might occupy isolated and very different environments through long-distance colonization or because of relictual distributions in refugia. Peripheral environments may prove important for long-term dispersal of the species and for evolution of unique morphs, subspecies, new species lineages, or entire species complexes. Often it is important to study the spatial and temporal scale of isolation, as well as the causes, in order to determine appropriate management actions that will maintain the condition or ameliorate undesirable isolation problems caused by human activities.

Likewise, fragmentation of environments in landscapes is not always undesirable. Remember that at some scale nearly all environments and species-specific habitats are fragmented. Environments or resources are often fragmented naturally through time: consider the seasonality of seeds and fruits, foliage, arthropods, and other items needed by animals. In some circumstances, natural fragmentation of environments or resources can lead to the evolution of new morphs or life forms. As is true of isolation, it is important to determine the causes and effects of fragmentation of environments to guide management action.

What should be done about isolation or fragmentation of habitats caused by human activity? The simple answer is to provide habitat linkages, including corridors or dispersed environments or resources, and to block habitats in the future. But these are blanket solutions that do not always meet a multiplicity of conservation objectives or fit the capability of the land or ownership patterns.

The Value of Remnant Patches

Citing potentially adverse effects of isolation and fragmentation should not be taken to mean that remnant patches of natural environments have no conservation value. Remnant patches may be all that is left in a landscape, watershed, or geographic region. They may be all we have from which to rebuild a more natural biota or ecosystem. Thus remnants may have much greater conservation value per unit area than do large natural sites, depending on the landscape context and management needs.

Remnant old-growth forests may be the last bastion for species closely associated with such environments—not only vertebrates but species of fungi, lichens, bryophytes, vascular plants, and invertebrates as well. Protection of small, isolated remnants may be worthwhile. In some situations, such conservation measures may be highly efficient in that they entail only a small land area but offer great benefits. Although small remnants would not support all life history requirements of mid- and large-size species such as wide-ranging carnivores, they can nonetheless provide valuable source pools for propagules and inocula of many plants and small animals vital to the function of native ecosystems and soil productivity (Amaranthus et al. 1994). And forest fragments provide at least some resources used by native vertebrates. Small remnants of native vegetation can be maintained for their value as environmental benchmarks, for scenic interest, or as sites of rich floral or faunal diversity.

Remnant patches of native environments perform a vital service by conserving sources of associated plants and animals and providing stepping-stone connectivity of a habitat for a species throughout a landscape. Beyond this, native remnants serve a purely utilitarian and anthropocentric need by supplying sources of valuable plants, pharmaceuticals, and foods (Schelhas 1995). Remnants also offer valuable learning experiences for restoration management.

Lessons

Conservation biologists have expended considerable time and effort developing concepts that might promote species persistence. Although the concepts make intuitive sense, remember that they are just concepts and usually have little empirical support—in large part because of the difficulty of replicating even small reserves and the multitude of interacting environmental factors.

Thus restoration projects that implement ecological and conservation concepts should be viewed as tests of these ideas. If designed properly (see Chapter 4), such projects can provide invaluable guidance for the future. This is why the publication of project results—regardless of the project's success—is such an essential part of the field of wildlife-habitat restoration.

Acknowledgments

This chapter was developed in collaboration with Thomas Scott from the University of California, Berkeley and Riverside.

References

Amaranthus, M., J. M. Trappe, L. Bednar, and D. Arthur. 1994. Hypogeous fungal production in mature Douglas-fir forest fragments and surrounding plantations and its relation to coarse woody debris and animal mycophagy. *Canadian Journal of Forest Research* 24:2157–2165.

Angelstam, P. 1996. The ghost of forest past—natural disturbance regimes as a basis for reconstruction of biologically diverse forests in Europe. Pages 287–337 in R. M. DeGraaf and R. I. Miller (eds.), *Conservation of Faunal Diversity in Forested Landscapes.* London: Chapman & Hall.

Angermeier, P. L., and J. R. Karr. 1994. Biological integrity versus biological diversity as policy directives. *BioScience* 44:690–697.

Austin, M. P., and C. R. Margules. 1986. Assessing representativeness. Pages 46–67 in M. B. Usher (ed.), *Wildlife Conservation Evaluation.* London: Chapman & Hall.

Beier, P. 1993. Determining minimum habitat areas and habitat corridors for cougars. *Conservation Biology* 7:94–108.

Beier, P., and R. F. Noss. 1998. Do habitat corridors provide connectivity? *Conservation Biology* 12:1241–1252.

Bennett, A. F. 1990. Habitat corridors and the conservation of small mammals in a fragmented forest environment. *Landscape Ecology* 4:109–122.

Bissonette, J. A. (ed.). 1997. *Wildlife and Landscape Ecology: Effects of Pattern and Scale.* New York: Springer-Verlag.

Bolger, D. T., T. A. Scott, and J. R. Rotenberry. 1997. Breeding bird abundance in an urbanizing landscape in coastal southern California. *Conservation Biology* 11:406–421.

Caughley, G., and A. Gunn. 1996. *Conservation Biology in Theory and Practice.* Cambridge, Mass.: Blackwell Science.

Chasko, G. G., and J. E. Gates. 1982. Avian habitat suitability along a transmission-line corridor in an oak-hickory forest region. *Wildlife Monograph* 82:1–41.

Church, R. L., D. M. Stoms, and F. W. Davis. 1996. Reserve selection as a maximal covering location problem. *Biological Conservation* 76:105–112.

Davidson, C. 1998. Issues in measuring landscape fragmentation. *Wildlife Society Bulletin* 26:32–37.

Diefenbach, D. R., L. A. Baker, W. E. James, R. J. Warren, and M. J. Conroy. 1993. Reintroducing bobcats to Cumberland Island, Georgia. *Restoration Ecology* 1:241–247.

Downes, S. J., K. A. Handasyde, and M. A. Elgar. 1997. Variation in the use of corridors by introduced and native rodents in south-eastern Australia. *Biological Conservation* 82:379–383.

Flather, C. H., K. R. Wilson, D. J. Dean, and W. C. McComb. 1997. Identifying gaps in conservation networks: Of indicators and uncertainty in geographic-based analyses. *Ecological Applications* 7:531–542.

Frankel, O. H., and M. E. Soulé. 1981. *Conservation and Evolution.* Cambridge: Cambridge University Press.

Freemark, K. E., J. B. Dunning, S. J. Hejl, and J. R. Probst. 1995. A landscape ecology perspective for research, conservation, and management. Pages 381–427 in T. E. Martin and D. M. Finch (eds.), *Ecology and Management of Neotropical Migratory Birds: A Synthesis and Review of Critical Issues.* New York: Oxford University Press.

George, T. L., and S. Zack. 2001. Spatial and temporal considerations in restoring habitat for wildlife. *Restoration Ecology* 9:272–279.

Haas, C. A. 1995. Dispersal and use of corridors by birds in wooded patches on an agricultural landscape. *Conservation Biology* 9:845–854.

Hall, L. S., P. R. Krausman, and M. L. Morrison. 1997. The habitat concept and a plea for standard terminology. *Wildlife Society Bulletin* 25:173–182.

Herkert, J. R. 1994. The effects of habitat fragmentation on midwestern grassland bird communities. *Ecological Applications* 4:461–471.

Hess, G. R. 1994. Conservation corridors and contagious disease: A cautionary note. *Conservation Biology* 8:256–262.

Holling, C. S. 1978. *Adaptive Environmental Assessment and Management.* IIASA International Series on Applied Systems Analysis. New York: Wiley.

Hutto, R. L. 1985. Habitat selection by nonbreeding, migratory landbirds. Pages 455–476 in M. L. Cody (ed.), *Habitat Selection in Birds.* San Diego: Academic Press.

Huxel, G. R., and A. Hastings. 1999. Habitat loss, fragmentation, and restoration. *Restoration Ecology* 7:309–315.

Johnson, D. H. 1980. The comparison of usage and availability measurements for evaluating resource preference. *Ecology* 61:65–71.

Knight, R. L., and J. Y. Kawashima. 1993. Responses of raven and red-tailed hawk populations to linear right-of-ways. *Journal of Wildlife Management* 57:266–271.

Landres, P. B., J. Verner, and J. W. Thomas. 1988. Ecological uses of vertebrate indicator species: A critique. *Conservation Biology* 2:316–328.

Lindenmayer, D. B., and H. A. Nix. 1993. Ecological principles for the design of wildlife corridors. *Conservation Biology* 7:627–630.

Lockwood, M., D. G. Bos, and H. Glazebrook. 1997. Integrated protected area selection in Australian biogeographic regions. *Environmental Management* 21:395–404.

Loehle, C., and B. Li. 1996. Habitat destruction and the extinction debt revisited. *Ecological Applications* 6:784–789.

Marcot, B. G., L. K. Croft, J. F. Lehmkuhl, R. H. Naney, C. G. Niwa, W. R. Owen, and R. E. Sandquist. 1998. Macroecology, paleoecology, and ecological integrity of terrestrial species and communities of the interior Columbia River basin and portions of the Klamath and Great basins. General Technical Report PNW-GTR-410. Portland, Ore.: USDA Forest Service.

Margules, C. R., and M. B. Usher. 1981. Criteria used in assessing wildlife conservation potential: A review. *Biological Conservation* 21:79–109.

Margules, C. R., A. O. Nicholls, and R. L. Pressey. 1988. Selecting networks of reserves to maximize biological diversity. *Biological Conservation* 43:63–76.

Marzluff, J. M., and K. Ewing. 2001. Restoration of fragmented landscapes for the conservation of birds: A general framework and specific recommendations for urbanizing landscapes. *Restoration Ecology* 9:280–292.

Mech, S. G., and J. G. Hallett. 2001. Evaluating the effectiveness of corridors: A genetic approach. *Conservation Biology* 15:467–474.

Meffe, G. K., and C. R. Carroll. 1997. *Principles of Conservation Biology.* 2nd ed. Sunderland, Mass.: Sinauer Associates.

Millar, C. I., and W. J. Libby. 1991. Strategies for conserving clinal, ecotypic, and disjunct population diversity in widespread species. Pp. 149–170 in D. A. Falk and K. E. Holsinger (ed.), *Genetics and Conservation of Rare Plants.* New York: Oxford University Press.

Mladenoff, D. J., T. A. Sickley, and A. P. Wydeven. 1999. Predicting gray wolf landscape recolonization: Logistic regression models versus new field data. *Ecological Applications* 9:37–44.

Morrison, M. L., B. G. Marcot, and R. W. Mannan. 1992. *Wildlife-Habitat Relationships: Concepts and Applications.* Madison: University of Wisconsin Press.

———. 1998. *Wildlife-Habitat Relationships: Concepts and Applications.* 2nd ed. Madison: University of Wisconsin Press.

Newmark, W. D. 1986. Mammalian richness, colonization, and extinction in western North American national parks. Ph.D. dissertation, University of Michigan, Ann Arbor.

Noss, R. F., M. A. O'Connell, and D. M. Murphy. 1997. *The Science of Conservation Planning: Habitat Conservation Under the Endangered Species Act.* Washington, D.C.: Island Press.

Orians, G. H. 1993. Endangered at what level? *Ecological Applications* 3:206–208.

Pengergast, J. R., R. M. Quinn, and J. H. Lawton. 1999. The gaps between theory and practice in selecting nature reserves. *Conservation Biology* 13: 484–492.

Pressey, R. L., and R. M. Cowling. 2001. Reserve selection and algorithms and the real world. *Conservation Biology* 15:275–277.

Pressey, R. L., and L. Tully. 1994. The cost of ad hoc reservation: A case study in western New South Wales. *Australian Journal of Ecology* 19:375–384.

Pressey, R. L., I. R. Johnson, and P. D. Wilson. 1994. Shades of irreplaceability: Towards a measure of the contribution of sites to a reservation goal. *Biodiversity and Conservation* 3:242–262.

Roberts, L. 1988. Hard choices ahead on biodiversity. *Science* 241:1759–1761.

Robinson, S. K., S. I. Rothstein, M. C. Brittingham, L. J. Petit, and J. A. Grzybowski. 1995. Ecology and behavior of cowbirds and their impact on host populations. Pages 428–460 in T. E. Martin and D. M. Finch (eds.), *Ecology and Management of Neotropical Migratory Birds: A Synthesis and Review of Critical Issues.* New York: Oxford University Press.

Rossi, E., and M. Kuitumen. 1996. Ranking of habitats for the assessment of ecological impact in land use planning. *Biological Conservation* 77:227–234.

Ryti, R. T., and M. E. Gilpin. 1987. The comparative analysis of species occurrence patterns on archipelagos. *Oecologia* 73:282–287.

Schelhas, J. 1995. Conserving the biological and human benefits of forest remnants in the tropical landscape: Research needs and policy recommendations. Pages 53–56 in J. A. Bissonette and P. R. Krausman (eds.), *Integrating People and Wildlife for a Sustainable Future.* Bethesda, Md.: Wildlife Society.

Schonwall-Cox, C. M., S. M. Chambers, B. Macbryde, and W. L. Thomas. 1983. *Genetics and Conservation: A Reference for Managing Wild Animal and Plant Populations.* Menlo Park, Calif.: Benjamin/Cummings.

Scott, J. M., F. Davis, B. Csuti, R. F. Noss, B. Butterfield, C. Groves, H. Anderson, S. Caicco, F. D'Erchia, T. C. Edwards, et al. 1993. Gap analysis: A geographic approach to protection of biological diversity. *Wildlife Monographs* 123:1–41.

Seabrook, W. A., and E. B. Dettmann. 1996. Roads as activity corridors for cane toads in Australia. *Journal of Wildlife Management* 60:363–368.

Simberloff, D., and J. Cox. 1987. Consequences and costs of conservation corridors. *Conservation Biology* 1:63–71.

Simberloff, D., J. A. Farr, J. Cox, and D. W. Mehlman. 1992. Movement corridors: Conservation bargains or poor investments? *Conservation Biology* 6:493–504.

Soulé, M. E. 1991. Land use planning and wildlife maintenance: Guidelines for conserving in an urban landscape. *Journal of the American Planning Association* 57:313–323.

Soulé, M. E., and D. Simberloff. 1986. What do genetics and ecology tell us about the design of nature reserves? *Biological Conservation* 35:19–40.

Soulé, M. E., A. C. Alberts, and D. T. Bolger. 1992. The effects of habitat fragmentation on chaparral plants and vertebrates. *Oikos* 63:39–47.

Stoner, K. E. 1996. Prevalence and intensity of intestinal parasites in mantled howler monkeys (*Alouatta palliata*) in northeastern Costa Rica: Implications for conservation biology. *Conservation Biology* 10:539–546.

Usher, M. B. 1986. *Wildlife Conservation Evaluation.* London: Chapman & Hall.

Vander Haegen, W. M., and R. M. DeGraaf. 1996. Predation on artificial nests in forested riparian buffer strips. *Journal of Wildlife Management* 60:542–550.

Van Vuren, D. 1998. Mammalian dispersal and reserve design. Pages 369–393 in T. Caro (ed.), *Behavioral Ecology and Conservation Biology.* New York: Oxford University Press.

Wallmo, O. C. 1969. Response of deer to alternate-strip clearcutting of lodgepole pine and spruce-fir timber in Colorado. Research Note RM-141. Fort Collins, Colo.: USDA Forest Service.

Welsh, H. H. 1990. Relictual amphibians and old-growth forest. *Conservation Biology* 4:309–319.

Wiens, J. A. 1989. *The Ecology of Bird Communities.* Vol. 2, *Processes and Variations.* Cambridge: Cambridge University Press.

CHAPTER 8

Wildlife Restoration: A Synthesis

Two messages should be evident from even a casual reading of this book: no wildlife study should be initiated without careful planning, and all such studies must be conducted with statistical rigor. Wildlife-habitat restoration requires that the conceptual framework driving our sampling design and methods be clearly stated. All of us operate within some conceptual framework, but this is seldom recognized and rarely stated. Are we basing our goals on the adage that "some data are better than no data at all"? Are we restoring portions of the historic vegetative community and assuming this will be acceptable to the historic animal assemblage? Do we think that competition determines the species' composition and abundance? Are we assuming that prey abundance is a satisfactory surrogate for prey availability? Regardless of your conceptual framework, the rigor you bring to a study will in large part determine the results. As we have seen, sample size is not a trivial matter, and no single sampling method is adequate for quantifying the presence, abundance, and activity of the animals in a project area.

Thus no study should be attempted unless it can be done within a clear conceptual framework and with rigor sufficient to convince others of the validity of the results. Each restoration project can be thought of as yet another piece of the puzzle that we hope will advance both the preservation of wildlife and our knowledge of basic animal ecology. Thus monitoring is an essential component of every project.

Major Messages

A central focus of this book is the use of habitat by animals. The term *habitat*, however, is a concept and hence cannot be tested per se. Traditionally it is understood to indicate a place where an animal resides. But this definition is useless as a predictive tool because the term has numerous many similar, but not identical, definitions. Thus habitat is an umbrella under which we can quantify specific relationships between an animal and its surroundings.

Habitat has a spatial extent that is determined during a stated time period. Thus the physical area occupied by an animal can be described by the observer. The various factors we commonly recognize as components of habitat—cover, food, water, and such—are contained within this area. A functional relationship between resources and animal performance is assumed. Often the observer does not define a specific area or produces a user-defined area that is perhaps based on animal activity. Thus habitat is a convenient boundary for measuring vegetation, various other resources, and the environment. User-defined boundaries may be convenient, but they are likely artificial because the resources they contain will vary over time, of course, both in response to abiotic factors and as a result of their use by animals. The spatial extent of the habitat, and the precision at which measurements are taken, should be defined to improve communication among workers. Descriptions of habitat should consider the dynamic nature of the components. Terms such as microhabitat, mesohabitat, and macrohabitat characterize the continuous nature of the factors that can be measured within an animal's habitat. Here again, technical limitations prevent microhabitat from being described and measured over a large extent at large mapping scales. Macrohabitat usually includes measures such as canopy cover and tree density, whereas microhabitat includes shrub stem density and pebble cover.

To quantify habitat quality requires us to discover which relationships determine individual fecundity and survivorship at the appropriate scale. The strength and frequency of interactions between the individual and its environment define the performance of the animal (survival, fecundity) and are considered niche relationships (Morrison and Hall 2002). Thus we can draw boundaries around where the animal performs activities and interacts with the biotic and abiotic characteristics—boundaries that can then be called the spatial extent of the habitat. Within this spatial extent we can then define the spatial and temporal resolution of our observations.

FIGURE 8.1.
Restoring native trees, such as the cottonwoods depicted here, without consideration
for shrubs and other understory plants (lower photo) often results in a monotypic
stand of limited value to wildlife (upper photo). (Photos courtesy of Annalaura Averill-
Murray and Suellen Lynn.)

Habitat can certainly be used to develop general descriptors of the distribution of animals (Figure 8.1). We fail repeatedly to find commonalties in
"habitat" for most populations across space, however, because we are usually
missing the underlying mechanisms (size distribution of prey, forage nutri-
ents, competitive factors) determining occupancy, survival, and fecundity.
Habitat per se can provide only a limited explanation of an animal's ecology.
Other concepts—especially the niche—must be invoked if we are to under-

stand the mechanisms responsible for animal survival and fitness. As we have seen in many wildlife studies (Collins 1983; Mosher et al. 1986; review by Morrison et al. 1998), "high-quality habitat" varies in physical attributes for a species across its range because the term is not measuring mechanisms. Rather, our statistical models of habitat are at best analyzing surrogates of these mechanisms. Habitat is a useful concept for describing the physical area used by an animal and should retain its simplicity for ease of communication among scientists, managers, and the public.

Developing a Restoration Plan

From the standpoint of wildlife, restoration planning can be thought of as either top-down or bottom-up. Top-down approaches predominate: we describe the desired vegetative community and develop plans to attain that condition. The time this will take depends, in large part, on the condition of the soil and the availability and size of planting stock. For practical and budgetary reasons, the major components of the vegetative community are prioritized for planting and maintenance. Other components—especially the understory (and this applies to grasslands as well)—are often more numerous but seldom emphasized. Here again, practical and budgetary concerns usually prevail. Unless a particular animal species is the focus of the restoration project, wildlife occupy the site in a haphazard manner.

I am not criticizing these top-down restoration efforts. Rather, I want to suggest that in many situations the bottom-up approach to restoration is a viable alternative. In the bottom-up approach, the requirements of each desired wildlife species are provided. Thus the restoration plan is *assembled.* This is obviously a complex and multifaceted approach that demands knowledge of the key habitat components of each species as well as the constraints (predators, competitors) on the use of resources by the desired species.

Although they are beyond the scope of this book, the many studies on multivariate analysis of habitat use are a useful tool for determining overlaps among species in key habitat variables. Figure 8.2, for example, is a theoretical representation of how species can be portrayed using multivariate statistics. Each circle represents a species (or sex or age class within a species), and the size of each circle indicates the breadth of use of different variables. Figure 8.3 shows results from a field study in which the range of habitat use is depicted. The extensive overlap shows that many species share key habitat components, although several show a narrow breadth of habitat use (that is, they are specialists). This figure is typical of the multivariate presentations

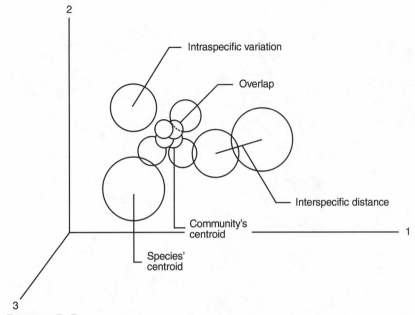

Model of a community in attribute space. The attributes could be morphology, diet, or other physiological, behavioral, or ecological parameters. The community is composed of some closely packed, invariable species and some distant, more variable kinds. (From J. S. Findley and H. Black, "Morphological and Dietary Structuring of Zambian Insectivorous Bat Community," Figure 1. *Ecology* 64:625–630. Copyright 1983.)

that are available. For a synthesis of habitat analysis that can be applied to restoration projects see Morrison et al (1998).

Providing key habitat components does not ensure that the species will occupy the site, of course, or successfully survive and reproduce. As we have seen throughout this book, many factors constrain the occupancy of a site and the activities that are possible. The population structure of a species will also influence its ability to colonize a site. Recall that many species have a metapopulation structure, and this has clear implications for restoration planning. Following a bottom-up approach to restoration planning cannot ensure that a project goal for a particular species will be met. You must evaluate the landscape context prior to initiating restoration. What are conditions like in the surrounding area? Is there a population of potential predators or competitors nearby? Are exotic species likely to invade the restored site and inhibit the target species from establishing?

Here are some of the key questions you must answer before proceeding with a restoration project:

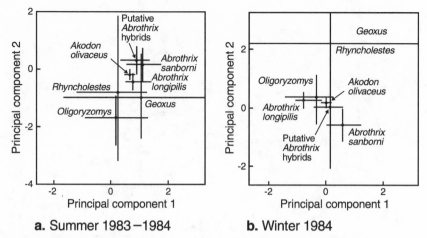

a. Summer 1983–1984   **b.** Winter 1984

FIGURE 8.3.
Results of principal-components analysis of habitat variables associated with capture sites of small mammals at La Picada, Chile. (a) Mean principal-component scores during summer. (b) mean principal-component scores during winter 1984. Projections are 95 percent CI. (From D. A. Kelt et al., "Quantitative Habitat Associations of Small Mammals in a Temperate Rainforest in Southern Chile: Empirical Patterns and the Importance of Ecological Scale," Figure 2. *Journal of Mammology* 75:890–904. Copyright 1994.)

- Are the critical factors present to allow establishment on the site? If not, can they be provided?
- Is the site large enough to support a viable population?
- Does the surrounding area (landscape context) present serious constraints (predators, competitors, human disturbance)?
- Can the constraints be managed over the short and long term to allow establishment and persistence on the site?
- Do linkages need to be developed from the site to other suitable sites to allow population persistence?

So far as I know, no restoration project has ever used a multispecies, bottom-up approach from design through implementation. My colleagues and I (Morrison et al. 1994a, 1994b) collected data on habitat use by herpetofauna, birds, and mammals to design a restoration plan for an urban park in southern California. Although we collected species-specific data, we did not use the information fully to build overlays of habitat use. We did, however, discuss the role that (primarily exotic) predators and feral animals could play in attaining a successful restoration of wildlife to the sites. Kus (1998) has reported on a useful method for guiding restoration for a single species, the endangered least Bell's vireo (*Vireo bellii pusillus*). She monitored vegetation structure at restoration sites and compared her results with a model of canopy architecture derived from vireo territories in natural areas. She also quantified the specific features of the sites (such as nest and foraging locations) that led to occupancy and nesting. This type of quantitative, comparative approach applies as well to multiple species on a single restoration site.

Information Gaps

Each chapter has pointed out the gaps in our information on animal ecology—weaknesses that inhibit our ability to develop rigorous restoration plans. The following topics deserve special attention because of their importance in developing restoration plans that center on wildlife habitat:

- We need better identification of the actual population boundaries of species of interest. This information has clear implications for understanding how individuals occupy a restored site. Thus studies of immigration, emigration, and dispersal are needed.
- We also need better determination of the metapopulation structure of a species of interest. Your project's success will be enhanced if you know the structure of animal populations throughout the region.
- We need to know more about the impact of competitors and predators on occupancy of a site. Extensive theoretical work has been done on how competitors and predators influence other species, but we have little practical data on how they influence occupancy of specific locations.
- Further work is needed to identify factors that limit the occupancy, survival, and reproductive success of animals. We can use these limiting factors—which are likely to include microhabitat variables and niche relationships—to build restoration plans.
- Little work has been conducted on the historic distribution and abundance of animals in a project area. Field notes and journals, museum records, and fossils and subfossils have been little utilized to design restoration plans.
- The influence of sampling effort and sampling methods on the results of field surveys needs much greater attention. Rare and abundant species usually call for different sampling methods and sampling intensities.
- We need to know more about the influence of the size, shape, and location of preserves on wildlife occupancy. Few empirical data are available to support or refute the numerous concepts on preserve design that have been offered.

- We need to consider the influence that the landscape surrounding each restoration project (the landscape matrix) could have on the specific animals of interest.
- Recent studies of corridors indicate that they function in a very species-specific and location-specific manner. We need more empirical evidence before we incorporate corridors into our restoration plans.

Moreover, we need to develop techniques for captive breeding, reintroduction, and translocation. Our understanding of the genetics of wild populations needs to be greatly expanded to aid in the design and management of captive populations. Recent work on the genetics of wild populations suggests that many subspecies' names do not predict groups with independent evolutionary histories—most likely because subspecies are usually based on arbitrary divisions of single character clines. Thus the subspecies we currently recognize are often misleading indicators of patterns of evolutionary history (Zink et al. 2001).

Working with Wildlife Scientists and Managers

Simple common sense tells us that people who work together should speak the same language. It thus behooves both restoration ecologists and wildlife scientists to become familiar each other's concepts, theory, and terminology. Take, for example, the different meanings of the term *habitat type* in plant and wildlife ecology. Additionally, there are multiple concepts of niche, community, the process of plant succession, and so forth. The message here is that restoration is a multifaceted endeavor that requires all workers to become knowledgeable in a host of ecological fields.

I suspect that many restorationists, especially trained plant ecologists, get frustrated with the myriad opinions among wildlife scientists regarding the appropriate variables to measure and the spatial and temporal scales at which the measurements should take place. As Morrison and Hall (2002) point out, even concepts with a long history in animal ecology are steeped in current controversy. The advent of remote sensing technology and GIS software has resulted in a substantial increase in large-area analyses of the distribution of wildlife. In fact, some workers declare that the landscape is the most effective scale for conserving wildlife (DeGraaf and Miller 1996). As we have seen throughout this book, I fundamentally disagree with this perspective—yet another demonstration that wildlife scientists often disagree on fundamental issues.

Such differences in perspective mean that restorationists must be current on fundamental issues of wildlife ecology. These issues are continually being discussed at professional meetings and debated in the pages of leading journals. Restorationists should participate in these discussions through membership in such organizations as The Wildlife Society, the Society for Conservation Biology, and the Ecological Society of America. In recognition of the central role that restoration can play in wildlife conservation, The Wildlife Society has developed a working group on restoration. Members of this group participate in focused discussions on issues pertinent to wildlife and wildlife-habitat restoration. Surely greater participation by restorationists in such organizations can only enhance the restoration of wildlife and their habitat.

Lessons

It is my hope that this book will aid in the restoration and management of wildlife. I consider this book a work in progress and therefore welcome feedback—both positive and negative—with the aim of creating a more useful edition in the future. Thus I invite all readers to contact me with suggestions for revision, including the addition of new or different material, modification of terminology, copies of pertinent references, and anything else that might improve this work. I am especially interested in hearing from students regarding the value of this material in the classroom. (As an instructor, I realize that my perception of course material is often rather different from that of my students.)

And lastly, I want to acknowledge all of the practitioners, managers, scientists, and students who are working so diligently in the service of wildlife and restoration. I hope this book will foster communication among all of us and lead to the growth of the field of restoration.

References

Collins, S. L. 1983. Geographic variation in habitat structure of the black-throated green warbler (*Dendroica virens*). *Auk* 100:382–389.

DeGraaf, R. M., and R. I. Miller (eds.). 1996. *Conservation of Faunal Diversity in Forested Landscapes*. New York: Chapman & Hall.

Findley, J. S., and H. Black. 1983. Morphological and dietary structuring of a Zambian insectivorous bat community. *Ecology* 64:625–630.

Kelt, D. A., P. L. Meserve, and B. K. Lang. 1994. Quantitative habitat associations of small mammals in a temperate rainforest in southern Chile: Empirical patterns and the importance of ecological scale. *Journal of Mammalogy* 75:890–904.

Kus, B. E. 1998. Use of restored riparian habitat by the endangered least Bell's vireo (*Vireo bellii pusillus*). *Restoration Ecology* 6:75–82.

Morrison, M. L., and L. S. Hall. 2002. Standard terminology: Toward a common language to advance ecological understanding and application. In J. M. Scott et al. (eds.), *Predicting Species Occurrences: Issues of Scale and Accuracy.* Washington, D.C.: Island Press.

Morrison, M. L., B. G. Marcot, and R. W. Mannan. 1998. *Wildlife-Habitat Relationships: Concepts and Applications.* 2nd ed. Madison: University of Wisconsin Press.

Morrison, M. L., T. A. Scott, and T. Tennant. 1994a. Wildlife-habitat restoration in an urban park in southern California. *Restoration Ecology* 2:17–30.

———. 1994b. Laying the foundation for a comprehensive program of restoration for wildlife habitat in a riparian floodplain. *Environmental Management* 18:939–955.

Mosher, J. A., K. Titus, and M. R. Fuller. 1986. Developing a practical model to predict nesting habitat of woodland hawks. Pages 31–35 in J. Verner, M. L. Morrison, and C. J. Ralph (eds.), *Wildlife 2000: Modeling Habitat Relationships of Terrestrial Vertebrates.* Madison: University of Wisconsin Press.

Zink, R. M., A. E. Kessen, T. V. Line, and R. C. Blackwell-Rago. 2001. Comparative phylogeography of some aridland bird species. *Condor* 103:1–10.

About the Author

Michael L. Morrison is manager of the University of California's White Mountain Research Station at Bishop. He obtained his Ph.D. in wildlife science from Oregon State University and has served on the faculties of the University of California–Berkeley, the University of Arizona, and California State University–Sacramento. He specializes in the study of wildlife/habitat relationships, restoration ecology, study design, and impact assessment. He is the author of several books on these topics, including *Wildlife-Habitat Relationships* (University of Wisconsin Press, 1998) and *Wildlife Study Design* (Springer-Verlag, 2001), and has written more than 150 scientific articles. He has designed and taught courses and workshops in wildlife habitat analysis, wildlife management, study design, and restoration ecology.

Index